AN INSIGHT INTO THE "KILLERS"- VIRUSES THAT CAUSE EPIDEMICS

AN INSIGHT INTO THE "KILLERS"-VIRUSES THAT CAUSE EPIDEMICS

Edited by:

Shiv Sanjeevi Sripathi
and Prerna Pandey

Delve
PUBLISHING

www.delvepublishing.com

An Insight into the "Killers"-Viruses that Cause Epidemics

Shiv Sanjeevi Sripathi and Prerna Pandey

Delve Publishing
224 Shoreacres Road
Burlington, ON L7L 2H2
Canada
www.delvepublishing.com
Email: orders@arclereducation.com

ISBN: 978-1-77407-805-1 (Hardcover)

DECLARATION

ABOUT THE EDITORS

ShivSanjeevi Sripathi completed his Masters in Biotechnology from Mumbai University in 2008. He was awarded for academic excellence in both his Bachelors and Masters for securing second rank in Mumbai University in 2006 and first rank in his college: Kishinchand chellaram College. For his Masters he secured first rank in his college KET's V.G.Vaze College. He qualified CSIR and NET and and TOEFL in September 2008. He then worked on a stem cell project at the Specialized Centre for Cell Based Therapy (SCCT),KEM Hospital at Mumbai on a project entitled, : Isolation & detection of stem cells from Human Umbilical cord/ amniotic membrane" following which he worked at Junior Research Fellow at Microbiology & Cell Biology Department, Indian Institute of Science, Bangalore from August 2009 to November 2011. The field of work involved cloning of cell wall genes and transcription factors in *E.coli* & *M.smegmatis*. As a writer, he has authored and co-authored 25 books with Delve and Arcler International Press. He loves to read and share on interesting aspects of life sciences in books. In his free time he loves to travel and explore ancient customs and traditions.

Dr Prerna Pandey, a biotechnologist with several years of wet lab research experience, has worked as a senior research fellow at International Center for Genetic Engineering and Biotechnology, New Delhi. Her field of research in PhD included isolation and molecular characterization of Geminiviruses, genome sequencing, gene annotation, as well as gene silencing using the RNA interference technology. She has published papers in peer-reviewed journals in the field and has a submitted a number of annotated Geminiviral genome sequences in the GenBank, including two novel ones. Prerna has also completed her editing and proof-reading courses from SfEP, UK and now works as a freelance scientific editor, writer and a subject matter expert. She has edited and co-authored several books on a range of topics in Life-Sciences. When Prerna is not busy with her assignments, she loves visiting historical places.

TABLE OF CONTENTS

LIST OF CONTRIBUTORS

Shiv Sanjeevi Sripathi

Prerna Pandey

Dustin Brisson
Department of Biology, University of Pennsylvania, Leidy Laboratories 209, 433 South University Avenue, Philadelphia, PA 19104-6018, USA

Catherine Brinkley
Department of Biology, University of Pennsylvania, Leidy Laboratories 209, 433 South University Avenue, Philadelphia, PA 19104-6018, USA

Parris T. Humphrey
Department of Biology, University of Pennsylvania, Leidy Laboratories 209, 433 South University Avenue, Philadelphia, PA 19104-6018, USA
Department of Ecology and Evolution, University of Arizona, BioSciences West room 310, 1041 E. Lowell St., Tucson, AZ, USA

Brian D. Kemps
Department of Biology, University of Pennsylvania, Leidy Laboratories 209, 433 South University Avenue, Philadelphia, PA 19104-6018, USA

Richard S. Ostfeld
Cary Institute of Ecosystem Studies, Box AB, Millbrook, NY 12545, USA

Roman Ullah
Department of Mathematics, Bacha Khan University, Charsadda 24420, Pakistan
Department of Mathematics, Abdul Wali Khan University Mardan, Mardan 23200, Pakistan

Gul Zaman
Department of Mathematics, University of Malakand, Chakdara 23101, Pakistan

Saeed Islam
Department of Mathematics, Abdul Wali Khan University Mardan, Mardan 23200, Pakistan

Jaber S. Alqahtani
UCL Respiratory, University College London, London, United Kingdom,
Department of Respiratory Care, Prince Sultan Military College of Health Sciences, Dammam, Saudi Arabia,

Tope Oyelade
UCL Institute for Liver and Digestive Health, London, United Kingdom,

Abdulelah M. Aldhahi
UCL Respiratory, University College London, London, United Kingdom,
Respiratory Care Department, Faculty of Applied Medical Sciences, Jazan University, Jazan, Saudi Arabia,

Saeed M. Alghamdi
National Heart and Lung Institute, Imperial College London, London, United Kingdom,
Faculty of Applied Medical Sciences, Umm Al-Qura University, Makkah, Saudi Arabia,

Mater Almehmadi
UCL Institute for Risk and Disaster, London, United Kingdom,

Abdullah S. Alqahtani
Department of Respiratory Care, Prince Sultan Military College of Health Sciences, Dammam, Saudi Arabia,

Shumonta Quaderi
UCL Respiratory, University College London, London, United Kingdom,

Swapna Mandal
UCL Respiratory, University College London, London, United Kingdom,
Royal Free London NHS Foundation Trust, London, United Kingdom

John R. Hurst
UCL Respiratory, University College London, London, United Kingdom,

Alexandra Teslya
Julius Center for Health Sciences and Primary Care, University Medical Center Utrecht, Utrecht University, Utrecht, The Netherlands

Thi Mui Pham
Julius Center for Health Sciences and Primary Care, University Medical Center Utrecht, Utrecht University, Utrecht, The Netherlands

Noortje G. Godijk
Julius Center for Health Sciences and Primary Care, University Medical Center Utrecht, Utrecht University, Utrecht, The Netherlands

Mirjam E. Kretzschmar
Julius Center for Health Sciences and Primary Care, University Medical Center Utrecht, Utrecht University, Utrecht, The Netherlands

Martin C. J. Bootsma
Julius Center for Health Sciences and Primary Care, University Medical Center Utrecht, Utrecht University, Utrecht, The Netherlands
Mathematical Institute, Utrecht University, Utrecht, The Netherlands

Ganna Rozhnova
BioISI—Biosystems & Integrative Sciences Institute, Faculdade de Ciências, Universidade de Lisboa, Lisboa, Portugal

Ayman Mubarak
Department of Botany and Microbiology, College of Science, King Saud University, Saudi Arabia

Wael Alturaiki
Department of Medical Laboratory Sciences, College of Applied Medical Sciences, Majmaah University, Majmaah 11952, Saudi Arabia

Maged Gomaa Hemida
Department of Microbiology and Parasitology, College of Veterinary Medicine, King Faisal University, Saudi Arabia
Department of Virology, Faculty of Veterinary Medicine, Kafresheikh University, Egypt

Quinton Katler
Department of Obstetrics and Gynecology, The George Washington University Hospital, Washington, DC, USA

Prachi Godiwala
Department of Obstetrics and Gynecology, The George Washington University Hospital, Washington, DC, USA

Charles Macri
Department of Obstetrics and Gynecology, The George Washington University Hospital, Washington, DC, USA

Beth Pineles
Department of Obstetrics, Gynecology and Reproductive Sciences, University of Maryland Medical Center, Baltimore, MD, USA

Gary Simon
Department of Internal Medicine, The George Washington University Hospital, Washington, DC, USA

Aileen Chang
Department of Internal Medicine, The George Washington University Hospital, Washington, DC, USA

Homa Ahmadzia
Department of Obstetrics and Gynecology, The George Washington University Hospital, Washington, DC, USA

Kathleen A. Alexander
Department of Fisheries and Wildlife Conservation, Virginia Tech, Blacksburg, Virginia, United States of America,

Claire E. Sanderson
Department of Fisheries and Wildlife Conservation, Virginia Tech, Blacksburg, Virginia, United States of America,

Madav Marathe
Department of Computer Science, Virginia Tech, Blacksburg, Virginia, United States of America,
Network Dynamics and Simulation Science Laboratory, Virginia Bioinformatics Institute, Virginia Tech, Blacksburg, Virginia, United States of America,

Bryan L. Lewis
Network Dynamics and Simulation Science Laboratory, Virginia Bioinformatics Institute, Virginia Tech, Blacksburg, Virginia, United States of America,

Caitlin M. Rivers
Network Dynamics and Simulation Science Laboratory, Virginia Bioinformatics Institute, Virginia Tech, Blacksburg, Virginia, United States of America,

Jeffrey Shaman
Department of Environmental Health Sciences, Mailman School of Public Health, Columbia University, New York, New York, United States of America,

John M. Drake
Odum School of Ecology, University of Georgia, Athens, Georgia, United States of America,

Eric Lofgren
Network Dynamics and Simulation Science Laboratory, Virginia Bioinformatics Institute, Virginia Tech, Blacksburg, Virginia, United States of America,

Virginia M. Dato
Department of Biomedical Informatics, School of Medicine, University of Pittsburgh, Pittsburgh, Pennsylvania, United States of America,

Marisa C. Eisenberg
Departments of Epidemiology and Mathematics, University of Michigan, Ann Arbor, Michigan, United States of America,

Stephen Eubank
Network Dynamics and Simulation Science Laboratory, Virginia Bioinformatics Institute, Virginia Tech, Blacksburg, Virginia, United States of America,

LIST OF ABBREVIATIONS

ACE2	Angiotensin-converting enzyme 2
ADE	Antibody-dependent enhancement
ADE	Antibody-mediated disease enhancement
ADEM	Acute Disseminated Encephalomyelitis
AIVs	Avian influenza viruses
APCs	Antigen-presenting cells
ARDS	Acute respiratory distress syndrome
CDC	Centers for Disease Control and Prevention
CDC	Chikungunya virus
COPD	Chronic obstructive pulmonary disease
CYD-TDV	Tetravalent dengue vaccine
DCs	Dendritic cells
DENV	Dengue virus
DF	Dengue fever
DHF	Dengue hemorrhagic fever
DPP4	Dipeptidyl peptidase-4
DRC	Democratic Republic of the Congo
dsRNAs	Double-stranded RNAs
EBOV	Zaire ebolavirus
FAO	Food and Agriculture Organization
FL	Fusion loop
HBV	Hepatitis B virus
HCV	Hepatitis C virus
HIV	Human immunodeficiency virus
HPAI	Highly pathogenic AI
HPAIVs	Highly pathogenic avian influenza viruses
ICU	Intensive care unit

KAP	Knowledge, attitudes, and practice
LPAI	Lowly pathogenic AI
MAPK	Mitogen-activated protein kinase
MERS	Middle East Respiratory Syndrome
MERS-CoV	Middle East respiratory syndrome coronavirus
Mo-DCs	Monocyte-derived dendritic cells
MVA	vaccinia virus Ankara
NAT	Nucleic acid testing
NPIs	Non-pharmaceutical interventions
PHEIC	Public Health Emergency of International Concern
PI3Ks	phosphoinositide 3-kinases
PMD	Para-menthane-diol
PRISMA	Preferred Reporting in Systematic Reviews and Meta-Analyses
PRNT	Plaque reduction neutralization test
PRNT	Plaque-reduction neutralization testing
PRRs	Pattern recognition receptors
RBD	Receptor binding domain
RIGI-	Retinoic acid inducible gene-I-
RNA	Ribonucleic acid
RT-PCR	Reverse transcriptase PCR
SARS	Severe Acute Respiratory Syndrome
SARS-CoV-2	Acute respiratory syndrome coronavirus 2
STRIVE	Sierra Leone Trial to Introduce a Vaccine against Ebola
SUDV	Sudan ebolavirus
TAFV	Taï Forest ebolavirus
TIR-	Toll/interleukin-1 receptor-
TLRs	Toll-like receptors
TMPRSS2	Tran's membrane serine protease
UAE	United Arab Emirates
UN	United Nations
US	United States
WHO	World Health Organization
WNV	West Nile virus
WWI	World War I

PREFACE

“There are only patterns, patterns on top of patterns, patterns that affect other patterns. Patterns hidden by patterns. Patterns within patterns. If you watch close, history does nothing but repeat itself”-Chuck Palahniuk.

This saying resonates with the occurrences of viral epidemics that strike the world from time to time. While history says the earliest recorded the influenza epidemic in 1500; we have heard of epidemics throughout history. Most of these viral epidemics like the 1918 Spanish Flu to the 2002-03 SARS epidemic. Let us not forget the H1N1, chikungunya virus (CHIKV), henipaviral diseases, Lassa fever, severe acute respiratory syndrome (SARS) virus, and the Middle East Respiratory Syndrome Coronavirus (MERS-Cov), Zika and Ebola virus epidemics. While most of the world was celebrating New Year’s Eve, 31st December 2019 reported the occurrence of a viral disease that spanned over most of 2020 to become the COVID-19 epidemic. Apart from the loss of lives, the virus has altered the economy and the way of life!

This book is the first of its kind: open its pages and delve into the world of the living and non-living and their effects.

This book discusses the “killer” viruses in terms of their transmission, control, and prevention. For a biologist, this is one concise book that presents data on the go to get an idea of the major epidemics and the viruses behind them. For a general reader, the book will open up a world of viruses-life forms considered neither living nor non-living yet have caused periodic deaths and lockdown of one of the most evolved forms of life-man!

An important point here is the points given under symptoms and treatment cannot be used for diagnosis and therapeutics. They are to enable the reader to understand what changes the virus brings as awareness is the first step towards preventing such viral diseases. Interesting aspects have also been presented such as the property of the influenza virus to undergo assortment and create new viral particles that go on to cause the next epidemic.

Most of these viral diseases can be prevented, making the saying “prevention is better than cure” hold true. While the search for vaccines is ongoing, the dengue virus and Zika virus can be prevented by controlling mosquito vectors while other viruses such as Coronavirus and influenza virus can be prevented by following hygiene practices and adhering to the guidelines set forth by the governments.

Incidentally, the book has been presented at the time when the COVID-19 epidemic is on. A vital point here is the requirement for cooperation among all humans. At such delicate times, rather than dividing and falling, we can stand together and overcome epidemics. This book also presents current research concerning the treatment and prevention of viral diseases to add value and show that hope should not be lost EVER.

1

An Introduction to Viral Epidemics and How they Can be Handled

Shivsanjeevi Sripathi

ShivSanjeevi Sripathi completed his Masters in Biotechnology from Mumbai University in 2008. He was awarded for academic excellence in both his Bachelors and Masters for securing second rank in Mumbai University in 2006 and first rank in his college: Kishinchand chellaram College. For his Masters he secured first rank in his college KET's V.G.Vaze College. He qualified CSIR and NET and and TOEFL in September 2008. He then worked on a stem cell project at the Specialized Centre for Cell Based Therapy (SCCT),KEM Hospital at Mumbai on a project entitled, : Isolation & detection of stem cells from Human Umbilical cord/ amniotic membrane" following which he worked at Junior Research Fellow at Microbiology & Cell Biology Department, Indian Institute of Science, Bangalore from August 2009 to November 2011. The field of work involved cloning of cell wall genes and transcription factors in *E.coli & M.smegmatis*. As a writer, he has authored and co authored 25 books with Delve and Arcler International Press. He loves to read and share on interesting aspects of life sciences in books. In his free time he loves to travel and explore ancient customs and traditions.

1.1 INTRODUCTION

Several factors of today considered as advancements contribute to the occurrences of epidemics such as climate change, changes in ecology, urbanization, and associated increases in travel. Epidemics, especially due to viruses are seeing a rise that places global health at risk.

While human health is threatened by infectious disease outbreaks, one cannot but help notice a rise in their occurrences. The 1918 Spanish influenza that ravaged the world had no reported history. There was an outbreak of yellow fever in the 1980s and also in 2015–2016 that affected Angola and the Democratic Republic of the Congo. The new century began with the 2002-03 SARS (Severe Acute Respiratory Syndrome) epidemic followed by the Ebola epidemic and the Zika virus epidemic. The years 2013 to 2016 saw an outbreak of the Ebola virus in Guinea, Sierra Leone, and Liberia. Another epidemic was due to the chikungunya virus (CHIKV), henipaviral diseases, Lassa fever, severe acute respiratory syndrome (SARS) virus, and the Middle East Respiratory Syndrome Coronavirus (MERS-Cov).

Of late, much in the news is the Coronavirus epidemic that has occupied prime news and thoughts of people across the globe. The entire world has been brought to a standstill due to this latest epidemic, not to mention the global death toll and morbidity!

Challenges in understanding many viruses and hence, design effective treatments have resulted in many curbs across the world: for example, in order to check the spread of the H5N1 avian flu, millions of domestic and wild birds were culled. Travel and tourism were brought to a halt due to the SARS epidemic in South East Asian countries and Africa on account of the Ebola virus. Due to the Zika virus, many Latin American countries have asked citizens to keep pregnancies on hold as the virus damages the nervous system of the fetus. The current COVID-19 virus epidemic has brought global tourism and trade to a screeching halt (Nandy and Basak, 2017; Kraemer et al, 2019).

1.2 WHY VIRAL EPIDEMICS

There are several factors that have a fundamental effect on the states of infectious diseases which in turn make it more difficult for prediction, and these factors are excessive population growth owing to rapidly altering demographics, and leading to deforestation, rampant urbanization, globalization of trade and universality of travel and of course, climate change. The ecology of existing vectors, the increase and rampant spread of drug-resistant infections and the ever-increasing vulnerability of large populations to sudden zoonotic disease outbreaks, like MERS (Middle East Respiratory Syndrome), Nipah, Ebola and so on, the ecology shows massive shifts, for instance, the explosion of the range of the Aedes mosquitoes, as well as the fact that the gamut of transmission is exacerbated by highly dense urban networks that are interconnected globally, especially for viruses like, Ebola, dengue, influenza and the SARS-CoV. It is when all these factors interject and combine that the incidences of complex epidemics get fueled.

To quote the WHO (World Health Organization) Health Emergencies Program Executive Director: "We are entering a very new phase of high-impact epidemics. This is a new normal, I don't expect the frequency of these events to reduce."

There are several communities that are the worst affected by epidemics and the priority should be to make these communities the focal point of the research by designating them as at-risk communities: since the local populace are usually the first responders and are the most driven to response, it is imperative that they are used in any form of response activities. The approach should be well integrated, despite the scope being local health agencies, communities that represent themselves as the regional health authorities and the international organizations as well as the national public health institutes. One of the most important roles is played by the World Health Organization, as they can use their unique mandate to alter faulty practices and facilitate the integration with global health practices while at the same time being respectful of the contributions of others.

One of the biggest challenges to working together in the dire situations of a viral outbreak is the vastly different basis of assumptions, varying cultures, work ethics and practices, each one individually well accepted within that particular community but maybe unacceptable elsewhere. But the crucial aspect that will foster success is in the understanding and acknowledging of the contribution of every party. The knowledge and awareness that in the toolbox that is ever-expanding, individual skills and knowledge bases are a very small part, has to be apparent to every participant. There is an urgent need to understand how to cohesively offer support in multi-stakeholder scenarios as well as the grassroots understanding that the actual path of major trends in research and innovation actually influences the pattern of an effective response to an epidemic (Bedford et al, 2019).

1.3 WE ARE INTERCONNECTED

This concept denotes the symbiosis between human and animal ecosystems and indicates as to how both are very tightly interconnected. This becomes important in the context of environmental changes- be it man-made or natural, bring about disruptions in the networks of pathogens, their hosts and the vectors in very complex ways, and lead to the decline as well as the emergence of an epidemic, endemic or zoonotic disease nearly impossible to predict. As animal diseases can hamper food production and access, it becomes a vital operation for One Health epidemic planning to monitor the pools of viruses, parasites, and bacteria that are resident in wild and domesticated animals. The US CDC or the United Nations (UN) Food and Agriculture Organization (FAO) allow for early identification of such zoonotic diseases to nip them in the bud (Bedford et al, 2019).

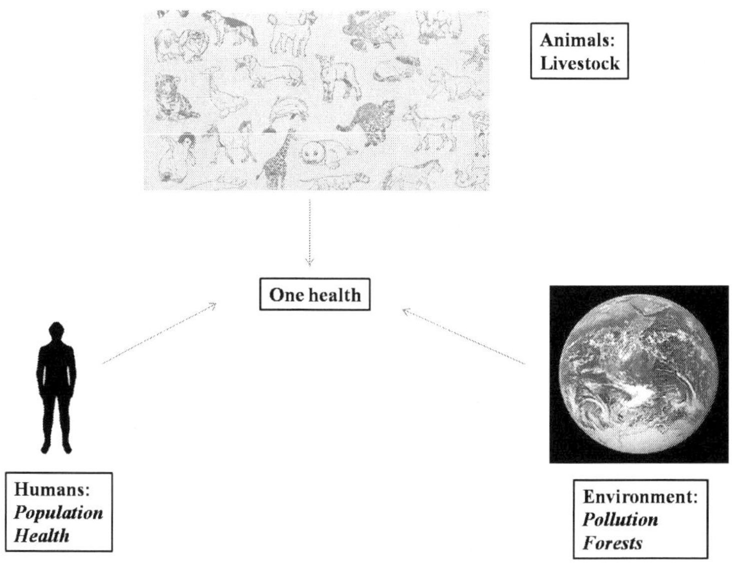

Figure 1.1. Interconnection between humans and the ecosystem.

1.4 RESPONDING TO VIRAL EPIDEMICS

Epidemics are a common enemy to all nations irrespective of borders. An epidemic takes out and disables health care, trust in authorities, routine health management, trade, travel, and vital commerce. This means that an internationally cohesive unified societal central initiative is key in assuming the broad risks and in the preparation and action response initiatives. There are the sprouts of seeds of such initiative starting to take root vis-à-vis coordinated initiatives. As an example, the Pandemic Influenza Preparedness Framework of the World Health Organization unifies nation-states, industry, other stakeholders along with the WHO to deploy a cohesive collected pan-global approach in the sphere of pandemic preparedness and response.

This brings with it a plethora of problems with regard to leadership, management and there needs to be a stringent oversight in order to minimize misconduct and issues in collaboration. Rampant or noticeable misuse and misappropriation of funds, manpower, and allied resources will lead to loss of goodwill among the people which in turn will prolong the epidemic/pandemic/outbreak.

An example of fake news and misinformation can be cited to show that high-income nations too are not immune to disruptions, due to distrust in Public health Systems and Public Authorities. That being said, it is imperative that the focus shifts from high-income nations so that there can be a median representation pan-global and the ability of deployment of the skills in the field in places where it is needed the most, as a result of successful global collaboration.

Regional Health Care Centers and National Disease Control Establishments should get total support with regard to expertise, governance, amicable sharing of resources, and capacity. Along with the principal task of protecting the population of the United States, the US-CDC (Center for Disease Control) offers its expertise all over the world. This spurred the European CDC to start offering its expertise in 2004, the Chinese CDC in 2015, and the African CDC in 2017. While there is no contesting the fact that there is more room for work to be done, concerning sharing of data as well as sharing of resources and facilities, the networking activities have indeed bolstered all their work and has also yielded in the cultivation of a very positive outlook towards these institutions in the eyes of the public from the low and medium-income nations.

In the 1830s, when the cholera epidemic swept across the continent, Europe witnessed several large scale riots: Medical equipment was pillaged and medical professionals like doctors, pharmacists, and nurses were murdered. Contemporarily, with regard to people who have not had prior interactions with Public Health Authorities, and are interacting with international teams only at the outset of an outbreak or pandemic, an immediate distrust for the mechanism and the intentions sets in and counterproductive activities start to happen in recovery and prevention initiatives against a "new" disease. Therefore, trust has to be earned on both sides. The community is the center point for all public health initiatives. Much before the outbreak, the outreach initiatives of the international teams have to reach the people at large and convey the message of the public health working for the benefit of the people, and the families of communities.

With the establishment of ground-level community monitoring of its residents, early detection of epidemics becomes a greater possibility. The knowledge base is first available only to the community when something happens—communication streams will help community volunteers identify occurrences and will have the tremendous advantage of being able to detect the diseases early and contain them before a spread.

This can be from several angles within the community, can be veterinarians who observe patterns in clusters of dead animals and birds, forgotten and side-cast workers in critical care facilities, nurses and doctors who are noticing patterns emerge in primary care; almost all of the major outbreaks like the H1N1, SARS, MERS, Nipah, H5N1 influenza subtype A were all initially observed by professionals in critical care facilities.

Another aspect is behavioral surveillance, which will be vital in responses to outbreaks of diseases in the twenty-first century. When we respond to the infectious disease outbreaks of the future, we will have to bear in mind that human behavior will continue to be very important. An example is just as how this surveillance was important in controlling the HIV epidemic to facilitate the identification and management of high-risk groups of the HIV infection. As an example several months went by before burial practices and transmission in hospitals were found to be the principal contributors to the rapid spread of the Ebola virus in West Africa in 2013.

Mobile phones, the internet, wireless internet, and social media are emerging technologies that show promise of being a very positive source for data in epidemiological studies, in assessing the mobility of populations in the case of the Ebola epidemic or the real-time data of the incidence of influenza. In the passage to precision public health, Machine Learning, Artificial Intelligence, and predictive technology will give tremendous boosts in the future as they post developments

1.4.1 Machine learning/AI (Artificial intelligence)

There have been numerous applications of Artificial Intelligence across spectrums of society thanks to the increase by leaps and bounds in speeds of computing as well as the tremendous progress in computer science. The existing process for detecting epidemics early is through sifting through online databases to study the symptoms of the diseases so that early detection is a possibility. Continued research can allow for AI to turn out reliable detections the techniques will have to be more sophisticated than what has been outlined (Shiv and Prerna, In press).

1.4.2 Other New Developments

In the case of Precision medicine, retracing the path of disease is performed using genomic sequencing in order to arrive at the best possible course of treatment for every individual. This idea by itself is very appealing in the eye of Public Health as it involves directly driving tailormade solutions for every individual.

The best demonstrator of the potential of this approach is in the two areas in the United States that posed a risk of Zika outbreak in 2016. The two areas measured were less than 5 km^2, and rather than involve the entire country, or even the state of Florida, all effort was focused on these specific neighborhoods. This can be contrasted to the "entire nations at risk" approach in 2016 in the case of Yellow fever (Dowell et al, 2016; Bedford et al, 2019).

Genomic epidemiology can monitor the evolution of a virus such as Ebola in an epidemic. It has been shown that these can detect outbreaks better than conventional epidemiology tools and hence is a vital addition to the arsenal. For instance, in Nigeria in 2018, the Lassa fever outbreak was proven to not be from a single Lassa virus variant, or human transmission using real-time genomic sequencing. However, the rodent reservoir barriers in migration as a result of major rivers across a vast geographical construct proved that there was a vast viral discrepancy among the nature of the virus (Siddle et al, 2018). The outbreak was contained thanks to these crucial findings.

Along with genomic information, the data can also include satellite imaging, mobile phone data, social media use data, and so on. For instance, in a study released in 2019, epidemiological surveillance data, travel surveys, parasite genetics were combined with anonymized mobile phone data to measure the spread of malaria parasites in southeast Bangladesh (Chang et al, 2019; Bedford et al, 2019).

1.4.3 Vaccines

There needs to be a wholehearted encouragement and support for research and development of vaccines. For example, at the time of the Ebola outbreak in West Africa in between 2013-2015, there were already pre-clinical data to support vaccine candidates that were actually ready for deployment, but there were a lot of challenges in deploying the clinical trials especially, considering that they needed to be done in a very time-sensitive manner in order to make available. However, in 2015, Zika was officially recognized as a pathogen that could lead to birth defects. Within a narrow span of two and a half years, not one, but three Phase One clinical trials were reported.

International systems are allowing for amalgamating the economics and logistics in terms of scaling up vaccine production in the case of a viral epidemic.

An example is the International Coordinating Group that maintains stockpiles of Yellow fever vaccines, oral vaccines for cholera, meningococcal vaccines and so on to prevent delays due to scaling up of manufacturing capacities during an epidemic

This also necessitates the need for pan-global initiatives with regards to future outbreaks, over several years. In the same vein, preclinical research should also be continual without hindrances in order to be ready for demand during future therapeutic pipelines and vaccines in the future (Bedford et al, 2019).

1.4.4 Ethic

The PREVENT project received Welcome funding in 2016 to develop and offer ethics guidance "at the intersection of pregnancy, vaccines, and emerging and re-emerging epidemic threats". This came in the footsteps of the new finding of the correlation between children being born with microcephaly and the occurrence of the Zika virus in pregnant mothers. While the development of a vaccine was the obvious route, researchers encountered a stumbling block where the risk to the unborn fetus was assumed to be greater than potential benefits as a result of testing on pregnant women. However, though the intention is noble there is a question of consent from the mother.

PREVENT was able to develop rapidly (Bedford et al, 2019) a proper guideline for the inclusion of pregnant women in the ongoing research for Zika virus and has since offered an extended scope for "a roadmap for the ethically responsible, socially just, and respectful inclusion of the interests of pregnant women in the development and deployment of vaccines against emerging pathogens."(Hume and Lua, 2017).

1.4.5 On the road to recovery

Also, all is not over with the end of a viral epidemic, as the survivors require care for a long time after. About three thousand children affected with microcephaly in Brazil are growing up who need special care. Microcephaly is one of the side effects of the Zika virus attack during pregnancy. Since most of these children will be growing up with very severe developmental issues, it is imperative that the tracking of these families and offering medical support will actually enhance our understanding of the Zika virus (Brickley and Rodrigues, 2018).

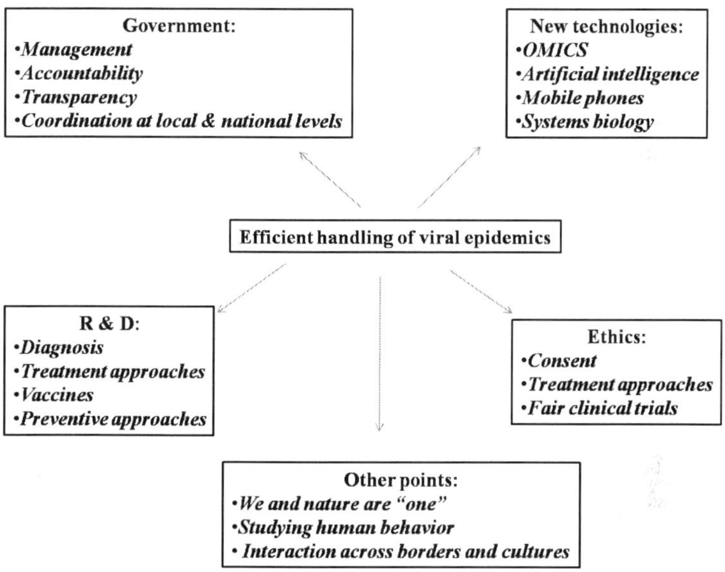

Figure 1.2. Schematics of various aspects that when integrated together can allow for efficient handling of viral epidemics.

1.5 CURRENT RESEARCH: A FEW INSIGHTS

1.5.1 Vaccines

The modes of transmission as well as actions of these viruses that cause epidemics are as different as cheese and chalk. This makes the development of approaches to control or cure them a challenge. It is observed that in order to any scientific system needs to begin at step one for each virus. Viruses are initially virulent for a period of time following which there is either a drop in their virulence or the development of immunity in people. For example, the World Health Organization (WHO) declared the Zika as Public Health Emergency of International Concern (PHEIC) in February 2016 but in November 2016, this term was removed and the virus was considered as a public health challenge by the WHO.

Additionally, the mutation rate of viruses (RNA viruses) is high that makes drugs developed against the initial strains of viruses ineffective in the later stages. For instance, oseltamivir (marketed as Tamiflu), considered effective against influenza shows resistant strains.

It is estimated that it takes 10-15 years and 1.8 billion $ to develop a drug from the lab to a patient!

The development of an effective drug molecule necessitates screening thousands of molecules and rigorous trials for the ones with promise. While containment-based approaches are just one side of the coin, another approach to address viral epidemics is vaccination. Vaccines are not new: Louis Pasteur brought in the successful idea of smallpox vaccination in the 19th century.

The major advantages in the use of vaccines are saving them time and development costs when compared with drugs (Chit et al, 2014). Vaccines allow one's own immune system to develop a response against the invading pathogens.

The development of vaccines in the face of an epidemic faces challenges in the form of transport, storing, and administering them across the affected area(s). New approaches are the need of the hour given that there is the factor of valuable time between identifying a pandemic agent and developing a vaccine.

The concept of vaccinomics refers to the use of immunology, genetics, and bioinformatics for efficient vaccine design. There is a wide range of options here: subunit vaccines or peptide vaccines to exploring the antigen-antibody interaction in the field of reverse vaccinology. This refers to tailor-made approaches for a particular community or population rather than one vaccine for the whole world.

Computerized tools are available to scan for antigenic segments on a virus to develop peptide vaccines. Examples include:

- IEDB (Immune Epitope Database) Analysis Resource (http://www.iedb.org/home_v3.php): can scan for and screen B-cell and T-cell epitopes

- PREDIVAC (http://predivac.biosci.uq.edu.au/): scans for HLA II binding
- ABCPred (http://crdd.osdd.net/raghava/abcpred/): predicts B-cell epitopes
- VaxiJen (http://www.ddg-pharmfac.net/vaxijen/VaxiJen/VaxiJen.html): can predict subunit vaccines

1.5.2 Modeling

The characteristics and the spatio-temporal dynamics of pathogens are described nowadays using empirical information and mathematical modeling.

This facilitates the guiding of surveillance, the resources required to tackle the disease, and the optimal use of resources.

Several aspects of epidemics have been inferred using phylogenetic and photodynamic tools. Genomic data yields information of the founder populations.

Such tools allow the study of how the disease was transmitted to new locations or even how to examine the risk of cross-transmission.

The comparison of epidemiological data with that of genomic data can allow for the quantification of incidences of the disease that may have been missed in an area and details inclusive of the epidemic doubling time and the basic reproductive number.

An example is the use of such analyses for the Zika Virus shown in figure 1.3 below in blue arrows.

Spatial information can be used to predict disease. This mapping allows explaining the area and landscape where the transmission occurs. Amalgamating this data with that of the travel patterns of humans can yield further information on the risk of disease, especially those carried by vectors such as Zika or Yellow fever or dengue.

In figure 1.3 below, green arrows show the estimation of geographical disease spreading.

As shown in (c) in the figure below, the estimates of geographic spread came out 6 months after the last case report that can , in turn, lead the surveillance program implementation.

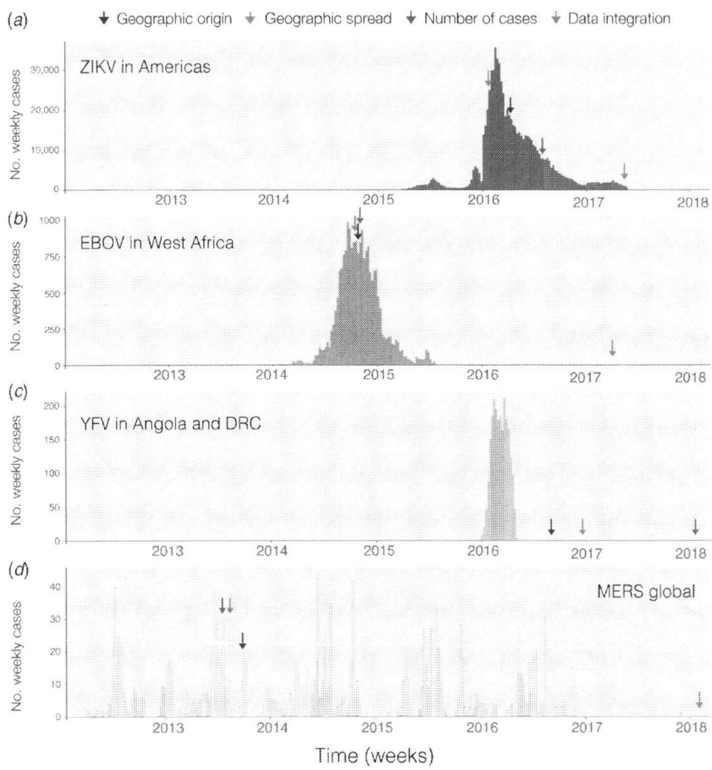

Figure 1.3. Modeling techniques to understand the dynamics of an epidemic (Kraemer et al, 2019).

Additionally, the features of how a disease is transmitted among humans also require to be understood. This commonly involves the reproduction number R (the average number of secondary infections generated by a case). In the case of a large MERS-CoV outbreak in South Korea (2015), the majority of infections were due to a small number of cases. Transmission chains can be reconstructed using genomic data.

Models can also guide the prediction of the possibility of an outbreak in the future in a given area. Coupled with real-time monitoring, there is an update in the predictions when a data point is received. An example is https://www.darpa.mil/news-events/2014-08-15 that predicted the chikungunya virus in the Americas.

The spread of a viral epidemic can be predicted using several data sources. A combination of epidemiological and genomic data can estimate the epidemic reproductive number. Data from spatial, genomic and epidemiological spheres have been combined by using latest technologies and interdisciplinary collaborations (as shown by red arrows in the figure above). This guided the introduction of factors contributing to the spread of Zika in the Americas. Spatial data coupled with phylogenetic analyses offer insights into the factors behind the spread of viruses.

Table 1.1. The amalgamation of genomics and epidemiology has allowed the monitoring of viral epidemics (Kraemer et al, 2019):

Characteristic of outbreak	Date published (online)	Citation
Zika virus outbreak in the Americas (2014-2017)		
Geographic origin	24 March 2016	Faria *et al.*, 2016 [2]
Geographic spread	14 January 2016	Bogoch *et al.*, 2016 [52]
Number of cases	25 July 2016	Perkins *et al.*, 2016 [23]
Integration of genomic/epidemiological/geographical data Ebola virus disease outbreak in West Africa (2013-2016)	17 May 2016	Grubaugh et aL, 2017 [33] and Faria et al., 2017[38]
Geographic origin	12 September 2016	Gire *et al.*, 2014 [53]
Geographic spread	2 September 2014	Gomes *et al.*, 2014 [54]
Number of cases	8 September 2016	Fisman *et al.*, 2014 [55]
Integration of genomic/epidemiological/geographical data	12 April 2017	Dudas *et al.*, 2017 [35]
Yellow fever outbreak in Central Africa (2015-2016)		
Geographic origin	20 September 2016	Grobbelaar *et al.*, 2016 [56]
Geographic spread	22 December 2016	Kraemer *et aL*, 2016 [3]
Number of cases	16 January 2018	Zhao *et al.*, 2018 [57]
Integration of genomic/epidemiological/geographical data	NA	NA
Middle East Respiratory Syndrome outbreak (2012-2017)		
Geographic origin	20 September 2013	Cotten *et al.*, 2013 [58]
Geographic spread	17 July 2013	Khan *et al.*, 2013 [59]
Number of cases	5 July 2013	Breban *et al.*, 2013 [60]
Integration of genomic/epidemiological/geographical data	16 January 2018	Dudas *et al.*, 2018 [42]

A 2016 published article screened 220 invertebrate species and discovered 1,445 RNA viruses, many of new families. They are capable of lateral gene transfer, evolution, and host switching (Shi et al, 2016). This makes the development of containment and treatment strategies to effectively combat viral diseases. An important aspect of any response to an epidemic is rapid response. An example is the Public Health Rapid Support Team to handle Ebola within 48 hours in any part of the world. Rapid and efficient research coupled with speedy delivery is the key to handling viruses that cause epidemics (Nandy and Basak, 2017).

1.6 FINAL POINTS:

The 21st century epidemics pose some real challenges. Poverty, socio-economic conflicts, irrational urbanization, adverse demographic transitions, and climate change are going to be palpably contributing factors to future epidemics. The two myths of epidemics need to be shattered, the first myth that they have to be handled discretely and second myth that they are isolated events. We need to dispel these myths and present a unified front even if we are not able to accurately predict the next upcoming outbreak. The right time, right volume and right intensity response strategy is then the name of the game. Every response to an epidemic yields several learning points and we need to build on those learning points and avoid negative outcomes in the subsequent ones.There are countries with very little resource mobilized for epidemiological study and research and therefore for a response as well, and it is fitting that these countries decide on their priorities at the outset. The problem with infectious diseases is that they have no sense of borders. The COVID-19 epidemic is such an example. There needs to be a sea change in the manner in which researchers and practitioners are being trained in all fields with respect to current epidemic scenarios. We need to pull down the artificial barriers between several disciplines, however disparate they seem and work beyond narrow lanes of epidemiological studies to multidisciplinary approaches to decipher how best to control and handle future epidemics.A positive change needs to be done to the entire gamut of the response infrastructure while a new generation of epidemiologists has to be trained inculcating the skillsets, knowledge base as well as networking capabilities so that they can readily access every tool available at their disposal. It is with this multi-disciplinary approach and a comprehensive and all-encompassing outlook that we can hope to defeat epidemics, protect lives irrespective of what sacrifices and cost it entails (Bedford et al, 2019).

1.7 REFERENCES

1. Nandy, A. and Basak, S.C. (2017). Viral epidemics and vaccine preparedness. *J Mol Pathol,* Epidemiol. 2: S1.
2. Kraemer, M.U.G., Cummings, D.A.T., Funk, S., Reiner, Jr.R.C., Faria, N.R., Pybus, O.G. and Cauchemez, S. (2019). Reconstruction and prediction of viral disease epidemics. *Epidemiol Infect*, 147, e34, 1–7. https://doi.org/10.1017/S0950268818002881
3. Bedford, J., Farrar, J. and Ihekweazu, C. et al. (2019). A new twenty-first century science for effective epidemic response. *Nature*, 575, 130–136. https://doi.org/10.1038/s41586-019-1717-y.
4. Gallagher, J. (7 June 2019). Large Ebola outbreaks new normal, says WHO. *BBC News*.
5. ShivSanjeevi, S. and Prerna P. (2020). Artificial Intelligence. Delve Publications. In Press.
6. Dowell, S. F., Blazes, D. and Desmond-Hellmann, S. (2016). Four steps to precision public health. *Nature*, 540, 189–191.
7. Siddle, K. J. et al. (2018). Genomic analysis of Lassa virus during an increase in cases in Nigeria in 2018. *N. Engl J Med*, 379, 1745–1753.
8. Chang, H.H. et al. (2019). Mapping imported malaria in Bangladesh using parasite genetic and human mobility data. *eLife*, 8, e43481.
9. Charlton, H.K. and Lua, L.H.L. (2017). Platform technologies for modern vaccine manufacturing. *Vaccine*, 35, 4480–4485.
10. Brickley, E. B. and Rodrigues, L. C. (2018). Further pieces of evidence in the Zika virus and microcephaly puzzle. *Lancet Child Adolesc Health*, 2, 162–164
11. Chit, A., Parker, J., Halperin, S.A., Papadimitropoulos, M. and Krahn, M., et al. (2014). Toward more specific and transparent research and development costs: The case of seasonal influenza vaccines. *Vaccine*, 32, 3336-3340.
12. Shi, M., Lin, X., Tian, J. et al. (2016). Redefining the invertebrate RNA virosphere. *Nature*, 540, 539–543. https://doi.org/10.1038/nature20167.

2

Influenza: The Historic and Current Pandemic

Shivsanjeevi Sripathi

ShivSanjeevi Sripathi completed his Masters in Biotechnology from Mumbai University in 2008. He was awarded for academic excellence in both his Bachelors and Masters for securing second rank in Mumbai University in 2006 and first rank in his college: Kishinchand chellaram College. For his Masters he secured first rank in his college KET's V.G.Vaze College. He qualified CSIR and NET and and TOEFL in September 2008. He then worked on a stem cell project at the Specialized Centre for Cell Based Therapy (SCCT),KEM Hospital at Mumbai on a project entitled, : Isolation & detection of stem cells from Human Umbilical cord/ amniotic membrane" following which he worked at Junior Research Fellow at Microbiology & Cell Biology Department, Indian Institute of Science, Bangalore from August 2009 to November 2011. The field of work involved cloning of cell wall genes and transcription factors in *E.coli* & *M.smegmatis*. As a writer, he has authored and co-authored 25 books with Delve and Arcler International Press. He loves to read and share on interesting aspects of life sciences in books. In his free time he loves to travel and explore ancient customs and traditions.

2.1 CONCEPTS AND OVERVIEW

The term Influenza is derived from the word: Influential or to influence in Latin. These are caused by three RNA viruses (A, B, and C) of the family Orthomyxoviridae and genus influenza. The types A and B cause both morbidity and deaths with the former showing a pandemic potential. There are annual epidemics of "seasonal influenza" on account of the high mutation rate of the genetic material of the virus. Between 28,000 and 111,500 deaths are reported in children less than 5 years across the globe due to lower respiratory infections associated with influenza (Saunders-Hastings and Krewski, 2016).

The subtypes of the influenza virus are differentiated on the basis of hem agglutinin (H) and neuraminidase (N) that are glycoprotein antigens. Of the identified 18 H subtypes and 11 N subtypes by the United States Centers for Disease Control and Prevention, human-to-human transmission is due to H1, H2, and H3 types.

These seasonal influenza spells are caused by mutations in the genes encoding the H and N antigens, known as antigenic drift that cause the virus to evade the immune system. Another system is the antigenic shift that is the formation of a new virus from the viral reassortment of two different influenza viruses. A novel influenza strain that caused pandemics included the H1N1 Spanish flu (1918), the H2N2 Asian flu (1957), the H3N2 Hong Kong flu (1968), and the H1N1 swine flu (2009).

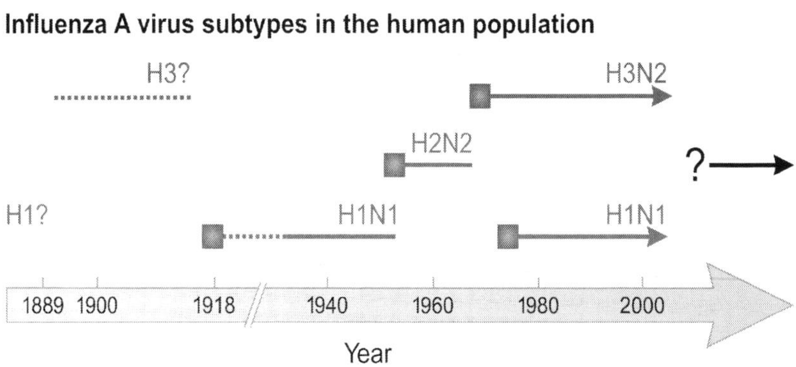

Figure 2.1. A chronology of the major influenza pandemics (*Source: https://en.wikipedia.org/wiki/Influenza_pandemic;* Saunders-Hastings and Krewski, 2016; Moghadami, 2017; Kebkiba et al, 2017).

Table 2.1. The features of influenza pandemics over the past 100 years

Strain	Pandemic	Years	Approximate mortality
H1N1	Spanish flu	1918–1920	40–50 million
H2N2	Asian flu	1957–1958	1–2 million
H3N2	Hong Kong flu	1968–1970	500,000–2 million
H1N1	Swine flu	2009–2010	575,000

An early influenza epidemic was described in Greek writings of 412 BC. Due to a lack of concrete data prior to 1500, the first influenza pandemic is considered to be one in 1580. This originated in Asia to North Africa then Europe and North America. In the 18th century, the first pandemic originated in 1729 in Russia to Europe, and the globe. The next one started in 1781 in China that later spread to Russia and Europe.

For the 19th century, the year 1830 saw a pandemic starting in China that was on the similar lines of the 1918 Spanish flu pandemic. However, the deaths were low though the illness rate was high. There was another one that originated in 1889 in Russia that spread to Europe and North America via sea and rail to cause the death of close to one million people across the world!Considered the 'greatest medical holocaust in history', the Spanish flu pandemic was caused by an H1N1 influenza strain. There were 3 waves: the spring of 1918, the fall of 1918, and the winter of 1918–1919 of which the second caused a huge loss of lives (40–50 million deaths). Apart from affecting so many people, there was a shutdown of business and education too. One interesting aspect of this pandemic strain was its high incidence in previously healthy young people (18 and 40 years of age).

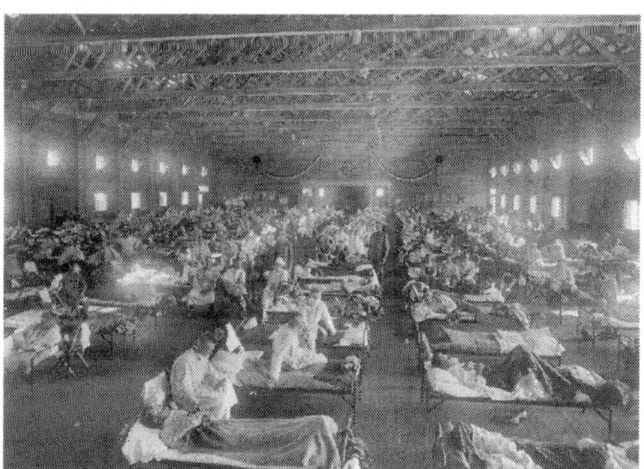

Figure 2.2. A grim scenario of the Spanish flu (*Source: https://en.wikipedia.org/wiki/Spanish_flu)*.

The Spanish flu pandemic is considered as a major health disaster given its effects on health and mortality. While populations would have not had contact with each other, World War I (WWI) brought countless people together in contact. This is a distinguishing factor of this pandemic given that trade routes and communication were the sources of spread of earlier pandemics while the conditions prevailing in trench warfare in Europe allowed for the transmission of the disease.

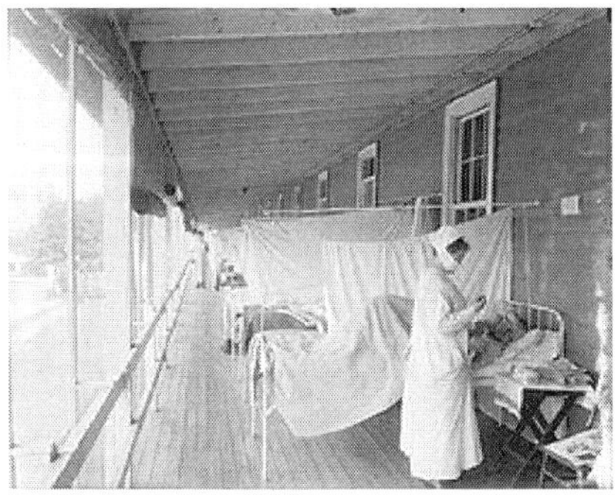

Figure 2.3. The Influenza ward at Walter Reed Hospital, in Washington, D.C. during the pandemic (*Source: https://en.wikipedia.org/wiki/Influenza_pandemic).*

As there were challenges in diagnosing the disease correctly and a shortage of medical staff, the outbreak was addressed by non-pharmaceutical interventions (NPIs). These ranged from closing schools, banning public gatherings, using face masks and following etiquette while coughing and sneezing. The following of such NPI practices lowered the peak of the pandemics and associated deaths, especially in the later waves of the pandemic.

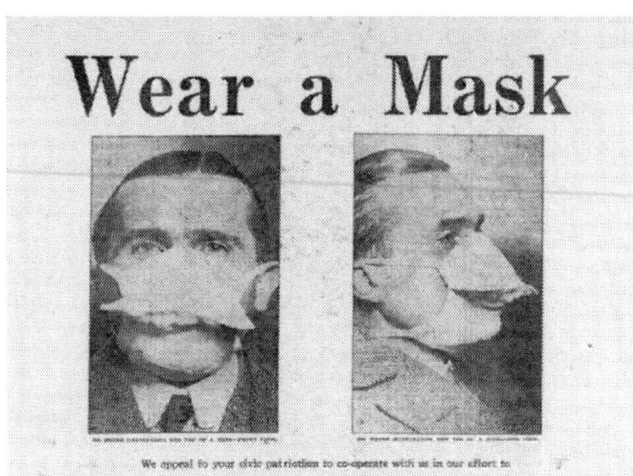

Figure 2.4. An advertisement published in 1918 an Oregon newspaper clip that reads: "We appeal to your civil patriotism to co-operate with us in our effort

to stamp out the Spanish Influenza or "Flu" Plague in Portland by wearing a mask." (University of Oregon).

Though the devastation brought by the Spanish flu was large: making it the worst pandemic throughout history, there were developments in medicine, international cooperation, and public health.

A new strain of influenza was detected in the Yunnan Province of China in 1957 that later spread to the whole world. This Asian flu caused the loss of between one and two million lives worldwide. This was the first pandemic that was studied in global surveillance systems and labs.

An antigenic shift in the Asian influenza virus produced the H3N2 that caused the Hong Kong flu. It was first reported in 1968 in Hong Kong and majorly targeted younger populations. On the lines of the Asian flu, this Hong Kong flu caused the loss of 500,000 and two million people across the world.

We now look at the Swine Flu caused by the pH1N1/09 virus in 2009 that was first reported in Mexico and the United States. The WHO declared it as a pandemic in June 2009 with its spread due to air travel and trade. The virus was the outcome of reassortment between 2 influenza viruses that targeted pigs. The exact effects of this pandemic are yet to be comprehended. There were high treatment costs coupled with a dip in the GDP by 0.5%–1.5% in affected countries, school closure and losses to the airline industry due to impaired tourism.

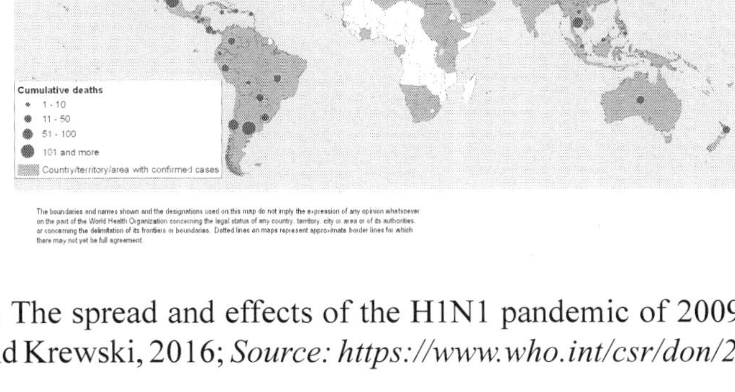

Figure 2.5. The spread and effects of the H1N1 pandemic of 2009 (Saunders-Hastings and Krewski, 2016; *Source: https://www.who.int/csr/don/2009_08_04/en/*).

2.2 THE VIRUSES THAT CAUSE THE FLU:

The term Influenza is derived from the word: Influential or to influence in Latin. These are caused by three RNA viruses (A, B and C) of the family Orthomyxoviridae and genus influenza. They can affect birds and mammals such as pigs, seals and humans. Type A viruses can infect many species but Types B and C affect humans with Type C also affecting pigs.

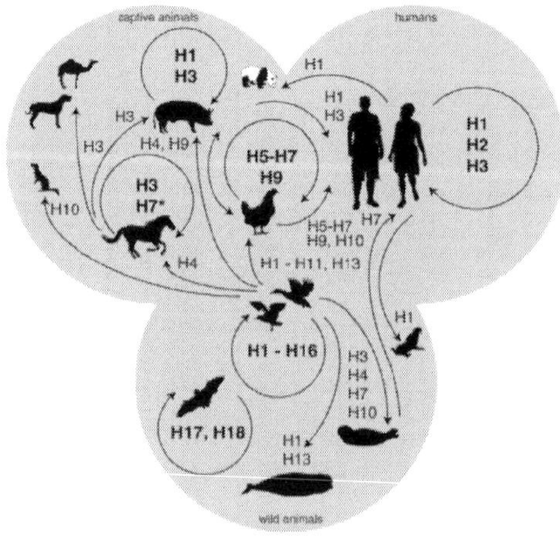

Figure 2.6. The transmission of the Influenza virus across species (Source: Short, Kirsty R et al. "One health, multiple challenges: The inter-species transmission of influenza A virus." One health (Amsterdam, Netherlands) vol. 1 (2015): 1-13. doi:10.1016/j.onehlt.2015.03.001).

Influenza viruses show the highest rate of attack in young people though the mortality is high in the elderly. Complications due to the disease are known to target the following populations:

- Infants (12-24 months) that are unvaccinated
- Women in the 2nd and 3rd trimesters of pregnancy
- Diabetes patients
- Patients with immunosuppressive disorders
- Patients with cognitive disorders
- Patients with cardiac disease
- Patients with cancer
- Patients with chronic kidney obstruction
- Patients with diseases involving abnormal hemoglobin
- Patients with lung diseases such as asthma or cystic fibrosis

The major source of the infection is large particle droplets from an influenza patient via coughing or sneezing. However, as these droplets are large and travel short distances, close contact is required for the disease to spread. Another source of spreading is touching contaminated surfaces (Moghadami, 2017).

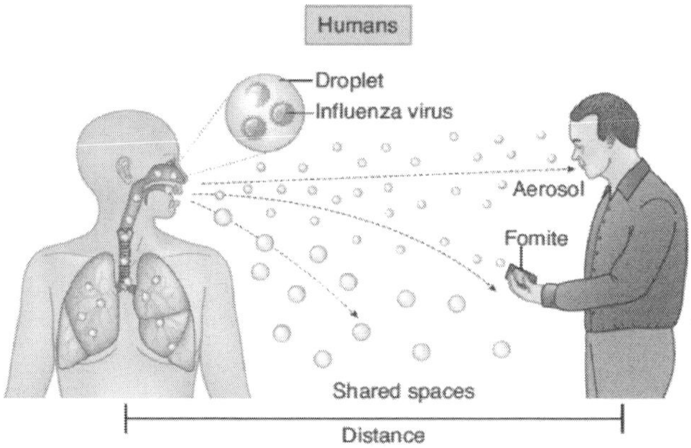

Figure 2.7. The spread of influenza virus (*Source: https://coryharow.wordpress.com/2015/11/18/influenza-virus-transmission-channels/).*

2.2.1 Symptoms

In the case of uncomplicated influenza, following 1-2 days of incubation, there is muscle pain (calf muscle pain especially in children) and even the back or eye muscles. There are routine symptoms of respiratory tract illnesses such as sore throat, nasal discharge, and dry cough. There is also a headache, chills, and fever.

The major complication of influenza is pneumonia:

Primary Influenza Viral Pneumonia:

Post-the run of the flu, there is fever, shortness of breath, and bluish skin and mucus membranes.

Secondary Bacterial Pneumonia:

The common causative agents are *Staphylococcus aureus* and *Streptococcus pneumonia* that cause wet cough, shortness of breath, and fever.

Other complications can include myocarditis and pericarditis that refer to inflammation of the heart muscle and heart lining, respectively.

2.2.2 Treatment

Treatment is deemed highly effective when the therapy is started in the first 24 hours. Either children or adults who belong to the "risk category" outlines above or those who require hospitalization should be administered treatment.

Patients with the H1N1 virus of the 2009 pandemic were mainly administered oseltamivir and zanamivir that are neuraminidase inhibitors. The administration of an NAI within 48 hrs after the appearance of the disease showed a lower chance of requiring ventilator support or dying.

Table 2.2. The recommended dosages and durations for treatment or prophylaxis (Moghadami, 2017):

Antiviral agent	Treatment	Details
Oseltamivir	Treatment	75 mg twice daily for 5 days
	Chemoprophylaxis	75 mg once daily for 7 days
Zanamivir	Treatment	10 mg (two 5-mg inhalations) daily for 5 days
	Chemoprophylaxis	10 mg (two 5-mg inhalations) once daily for 7 days

Other approaches to treating the disease have also been explored to overcome the limitations of resistance seen with anti-viral medications. These include:

- Monoclonal antibodies such as those aimed at the conserved M2 protein (M2e)
- Using corticosteroids: Here there are conflicting reports with some reports showing that corticosteroids were important during the 2009 influenza in Europe (Linko et al, 2011) while the administration of corticosteroids has been associated with higher mortality (Rodrigo et al, 2015).

2.2.3 Vaccines

The annual vaccination against seasonal influenza is considered vital to prevent influenza and its associated complications. As the virus is known for its high rate of mutation to evade the immune system, new vaccines are generated in accordance with the virus variants. Global surveillance data are used to guide the correct selection of antigens for designing vaccines.

In the southern hemisphere, recommendations are made in September for the next influenza season while it is in February for the northern hemisphere given the 6-8 month time frame for developing and approving vaccines (Moghadami, 2017).

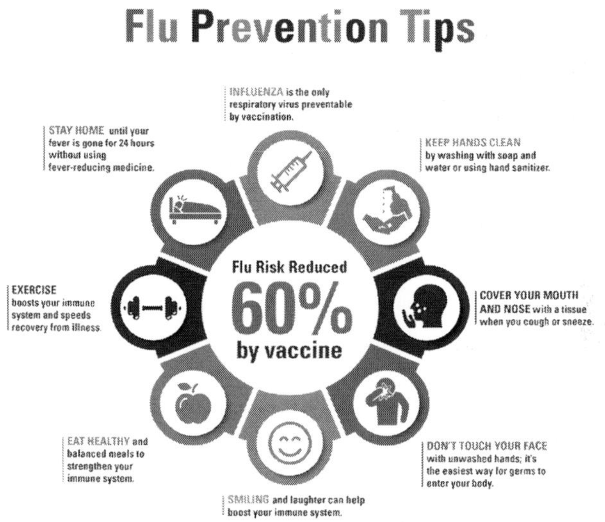

Figure 2.8. The approaches one can follow to prevent an attack by the influenza virus (*Source: https://www.nationaljewish.org/conditions/health-information/health-infographics/flu-prevention-tips)*.

2.3 AVIAN INFLUENZA: BRIDGING SWINE AND HUMANS

A group of natural viruses found in wild aquatic birds, such as ducks, swans, geese, shorebirds, gulls, and terns is avian influenza viruses (AIVs). Of these, the subtype H5N1 is considered as highly pathogenic avian influenza viruses (HPAIVs) was discovered back in 1996 in Guangdong province in China in geese. An outbreak resulted in a mortality rate of 40% in farmed geese: this spilled over to the live poultry markets of Hong Kong. This was coupled with 18 confirmed cases in humans of whom, 6 lost their lives. The transmission from birds to humans was confirmed by high homology between the isolates from birds and humans. The containment was done by the culling of poultry in Hong Kong. Novel H5N1 variants resulted in 2 humans getting infected in 2003. The year 2017 saw that 860 cases of H5N1 influenza virus infection resulting in 454 people losing their lives. As 64% of the s domesticated ducks and 95% of the domesticated geese populations in the world are in China, these viral outbreaks have negatively affected the poultry market and hence, economy (Nuñez and Ross, 2019).

Few viruses of avian origin can also extend their host range to cats, dogs, mice, ferrets, horses, and as mentioned above humans. The years 1999 to 2000 also saw an H5N1 outbreak in Italy in chicken and turkey farms.

These Type A influenza viruses are called highly pathogenic AI (HPAI) or Lowly pathogenic AI (LPAI) based on their ability to cause disease manifestations. While the H5 and H7 subtypes of the HPAI strains can cause 100% mortality in domesticated birds, the LPAI strains are usually not lethal with sometimes no symptoms in infected birds.

The main source of the virus is a sick bird, mainly through nasal discharge, eye secretions, and feces. In the case of aquatic wild bird reservoirs, the major source of infection is the fecal-oral transmission. The viruses can also be transmitted to mice that feed on the dead bodies of affected birds or flies or ticks and earthworms. Feces that are contaminated can be infectious for months. There is also the possibility of indirect contact through water or food or clothing or shoes or trucks and staff that are in contact with the birds. Closed spaces such as farms or markets cause close contact which, in turn, facilitates the spread of the virus.

The signs of the disease show variation given the virulence of the viruses. While LPAI can involve mild respiratory disease, HPAI cause lowered eating, respiratory distress, diarrhea, watery eyes, nervous disturbances, misshapen eggs, nasal discharge, and sometimes death without manifesting any symptoms.

Apart from the economic impact due to the affected poultry market, sometimes the public or governments may resort to hysterical measures such as uncontrolled culling and nest destruction. For example, H5N8 was seen in The Netherlands, Germany, Canada, East Asia, and the UK that resulted in the culling of 48 million chickens and turkeys amounting to a loss of US$1.6 billion!

Another point of concern is the fact that pigs serve as systems where viral reassortment occurs to generate new viruses. This can account for infection in humans. The respiratory tract of swine contains -2,3 and -2,6 sialic acid receptors to which the protein HA binds to allow for viral entry. There have been epidemics in pig populations with newly generated (assorted) H3N2 swine viruses.

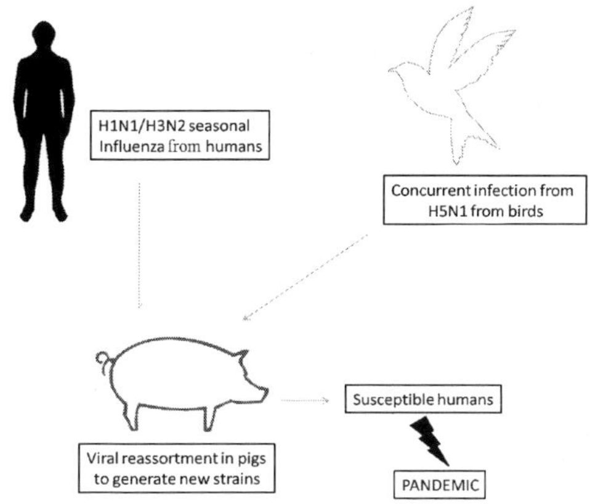

Figure 2.9. The transmission of influenza viruses among humans, birds and pigs that can result in viral reassortment and subsequent infection in humans leading to pandemics.

In the case of humans, infection due to HPAIV has been seen occasionally in Asia, Africa, the Middle East and Europe. There was a case of human H5N1 infection in the USA, Canada in 2014 in an individual who had been to China. The infection starts off as the typical uncomplicated influenza with respiratory irritation, body ache and fever. This can later develop as a lower respiratory tract infection, severe pneumonia, organ failure and septic shock.

There have been 454 deaths in the world due to HPAIVs (2018 data). The chances of human-human transmission are decreasing and are also challenging for the virus to do so. Though there is no immediate threat of such a pandemic, it must be kept in mind that transmission between mammals are just a few mutations away (Kebkiba et al, 2017, Nuñez and Ross, 2019).

2.4 FINAL POINTS

Given the swine flu and avian flu pandemics, it emerged that apart from humans, swine, birds and hence, animal populations too need global surveillance. This becomes important especially in areas where there is a significant overlap between humans and animals. Given that contact tracing for influenza becomes challenging and costly even for small populations, global cooperation between governments should be increased. A point is being reiterated that influenza viruses are capable of viral reassortment and what is termed as antigenic drift, as well as antigenic shift, that accounts for their constant mutational rates. The 3 viruses that caused pandemics: The three pandemic viruses that emerged in the 20th century — the 1918 ("Spanish influenza") H1N1 virus, the 1957 ("Asian influenza") H2N2 virus, and the 1968 ("Hong Kong influenza") H3N2 virus, as discussed have wreaked havoc. The exchange of genetic material between 2 influenza A viruses resulted in a new virus that was associated with the 1957 and 1968 pandemics. Double infection of a human and another animal (possibly pig) with avian H2N2 influenza and human H1N1 influenza virus produced a new virus in 1957 that contained 5 segments from the human virus and the hemagglutinin, the neuraminidase, and the gene for one of the polymerase proteins (PB1) from the avian virus. This virus was present in humans until 1968 where the H3N2 Hong Kong virus-a product of reassortment itself. This had avian genes in place of hemagglutinin (H2) and polymerase (PB1) genes of the H2N2 virus (Belshe, 2005).

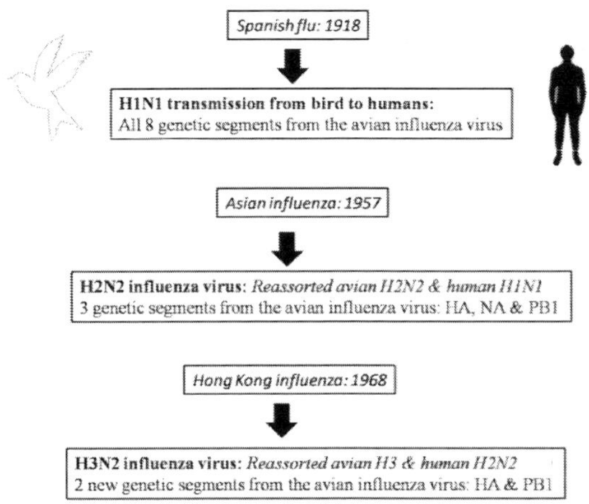

Figure 2.10. The reassortment in influenza viruses that have resulted in pandemics (Inspired by Belshe, 2005).

Thus, a streamlining of the disease reporting is vital along with shorter lag between reports and evaluation. The major cause of disease or death is when people do not avail treatment that hence, calls for population-based surveillance in real-time possibly with the use of social media. As viral pandemics and their patterns are uncertain, the responses and policies require both agility and flexibility to nip them in the bud. Scientific studies and global surveillance are integral to developing therapy and vaccines against such outbreaks and be ready for the next pandemic (Saunders-Hastings and Krewski, 2016; Nuñez and Ross, 2019).

REFERENCES

1. Saunders-Hastings, P. R., & Krewski, D. (2016). Reviewing the History of Pandemic Influenza: Understanding Patterns of Emergence and Transmission. Pathogens (Basel, Switzerland), 5(4), 66. https://doi.org/10.3390/pathogens5040066
2. Moghadami M. (2017). A Narrative Review of Influenza: A Seasonal and Pandemic Disease. Iranian Journal of Medical Sciences, 42(1), 2–13.
3. Kebkiba B, Antipas B-B, Mahamat O. (2017). Avian Influenza: Review. IOSR Journal of Agriculture and Veterinary Science (IOSR-JAVS) e-ISSN: 2319-2380, p-ISSN: 2319-2372: PP 34-40
4. Nuñez IA, Ross TM. (2019). A review of H5Nx avian influenza viruses. Therapeutic Advances in Vaccines and Immunotherapy, 7, 2515135518821625. https://doi.org/10.1177/2515135518821625.
5. Belshe RB. (2005) The Origins of Pandemic Influenza — Lessons from the 1918 Virus. The New England Journal of Medicine 353;21 www.nejm.org.

3

COVID 19-The virus that locked the world down

Shivsanjeevi Sripathi

ShivSanjeevi Sripathi completed his Masters in Biotechnology from Mumbai University in 2008. He was awarded for academic excellence in both his Bachelors and Masters for securing second rank in Mumbai University in 2006 and first rank in his college: Kishinchand chellaram College. For his Masters he secured first rank in his college KET's V.G.Vaze College. He qualified CSIR and NET and and TOEFL in September 2008. He then worked on a stem cell project at the Specialized Centre for Cell Based Therapy (SCCT),KEM Hospital at Mumbai on a project entitled, : Isolation & detection of stem cells from Human Umbilical cord/ amniotic membrane" following which he worked at Junior Research Fellow at Microbiology & Cell Biology Department, Indian Institute of Science, Bangalore from August 2009 to November 2011. The field of work involved cloning of cell wall genes and transcription factors in *E.coli* & *M.smegmatis*. As a writer, he has authored and co-authored 25 books with Delve and Arcler International Press. He loves to read and share on interesting aspects of life sciences in books. In his free time he loves to travel and explore ancient customs and traditions.

3.1 INTRODUCTION

The word "2019 novel coronavirus disease (COVID-19)" induces chills in the minds of global citizens who have lived in the 21st century. The loss of lives and restrictions in the economy has left an indelible mark that may take longer to be erased.

This infectious disease is caused by severe acute respiratory syndrome coronavirus 2 (SARS-CoV-2). Wuhan in the Hubei province of China was the first site where the disease that has become a pandemic altering the health scenario of the world was first reported. Adults in the city known to be a major transportation hub of China reported to hospitals with severe pneumonia whose cause was not known. Several of these cases reported a common exposure to the Huanan wholesale seafood market known for also selling live animals.

The Wuhan Municipal Health Commission on 31st Dec 2019 reported to the World Health Organization or WHO that this was unknown pneumonia that was being subject to close monitoring. On 1st January 2020, the market was closed. The identification that it was a coronavirus was on 7th January. This virus showed a similarity of 70% to the SARS-CoV and around 95% homology with bat coronavirus.

The cases then rose up exponentially showing the ability of the virus to transmit among humans. 11th Jan 2020 was the day of recording the first human death. This was further fueled by the Chinese New Year associated migration. Cases were reported from people returning from Wuhan to Thailand, Japan, and South Korea. This was followed by reports of the disease being transmitted to healthcare workers on 20th Jan 2020. A "lockdown" was declared in Wuhan that was extended to other cities in the province.

On 30th January 2020, the WHO declared the outbreak to be a Public Health Emergency of International Concern.

Transmission among humans was established when people in other countries developed COVID-19 despite not traveling to China. This led to screening operations in airports to test for COVID-19 and isolating the symptomatic people. It was soon found that the virus could also be spread prior to symptom onset or from asymptomatic people.

11th March 2020 was the day the disease was declared as a pandemic with the number of people diagnosed touching 1 million on April 2, 2020. The fatality rate was 5.2% across 204 countries as against the SARS epidemic of 2002-2003 that infected 8096 people in 29 countries and the Middle East respiratory syndrome coronavirus (MERS-CoV) that targeted 2494 people in 27 countries between 2012-2019.

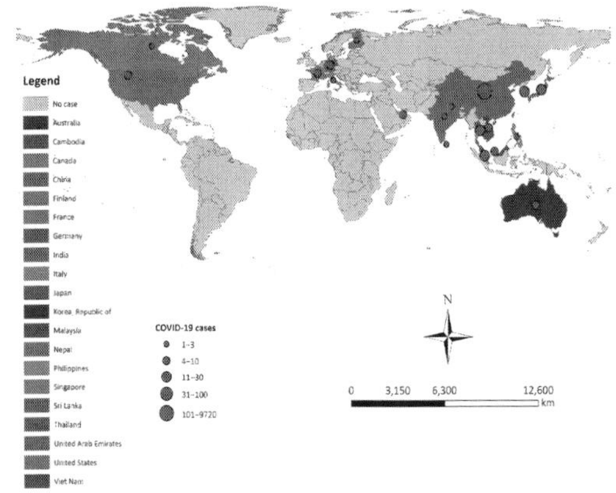

Figure 3.1. The COVID-19 pandemic during the early outbreak (Source: Adhikari, S., Meng, S., Wu, Y. et al. Epidemiology, causes, clinical manifestation and diagnosis, prevention and control of coronavirus disease (COVID-19) during the early outbreak period: a scoping review. Infect Dis Poverty 9, 29 (2020). https://doi.org/10.1186/s40249-020-00646-x).

As of July 1, 2020, more than 200 countries were affected by SARS-CoV-2 causing 10 million identified cases with 508 000 confirmed deaths (Chauhan, 2020; Singhal, 2020; Wiersinga et al, 2020).

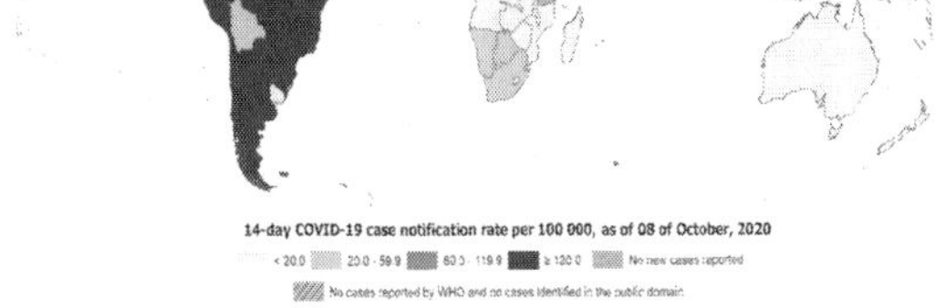

Figure 3.2. The Geographic distribution of reported COVID-19 cases per 100 000 population (14-day cumulative number, as of 5th August 2020; *Source: https://www.ecdc.europa.eu/en/geographical-distribution-2019-ncov-cases*).

3.2 THE VIRUS

Coronaviruses are positive-sense RNA viruses; 60 nm to 140 nm in diameter and are enveloped. The name coronavirus refers to their crown-like appearance under the electron microscope due to the spike- like projections on the surface (Singhal, 2020).

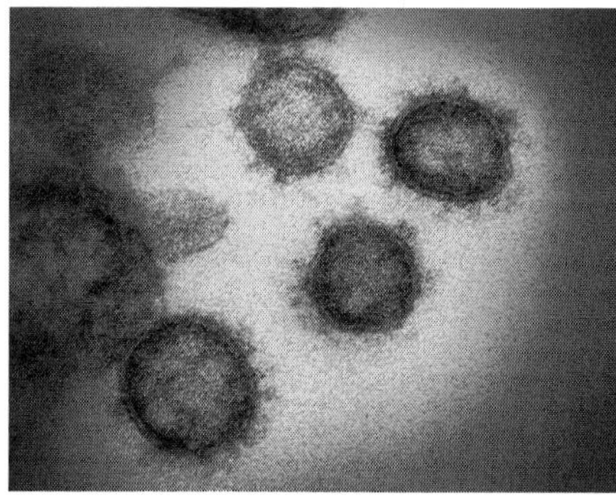

Figure 3.3. An electron microscope image of the COVI-19 coronavirus; notice the crown-like projections on the surface (Source: NIAID-RML; npr.org).

The coronaviruses belong to the family *Coronaviridae* and order *Nidovirales* infecting companion animals, livestock, and birds. Here, they cause respiratory, neurological, enteric, and cardiovascular diseases.

In humans, the viruses cause respiratory and gastrointestinal diseases ranging from pneumonia, bronchitis to severe acute respiratory distress syndrome (ARDS), multi-organ failure, and mortality. They have also been linked to asthma, cystic fibrosis, and chronic obstructive pulmonary disease (Zimmermann and Curtis, 2020).

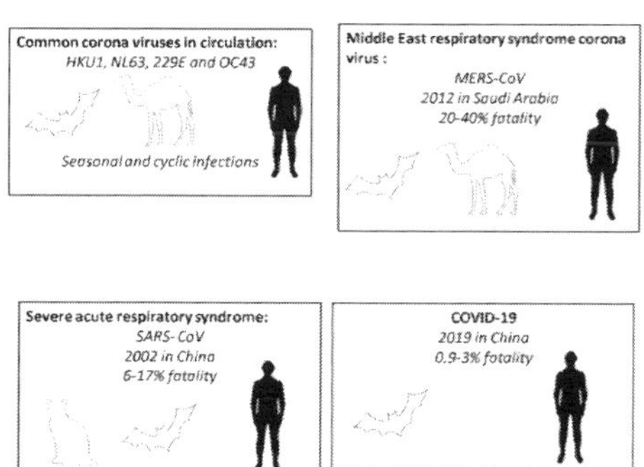

Figure 3.4. The summary of coronaviruses (Adapted from Zimmermann and Curtis, 2020).

In humans, HKU1, NL63, 229E, and OC43 are 4 coronaviruses known to be in circulation to cause mild respiratory diseases. Severe disease has been the outcome of crossovers between animal coronaviruses to humans.

There was an outbreak of severe acute respiratory syndrome coronavirus in 2002-2003 in the Guangdong province of China.

This virus was the result of the crossover of a β coronavirus from bats to humans via the palm civet cat as the intermediary host. 8422 people in mostly China were targeted with an 11% mortality rate.

n 2012, there was another crossover from a bat coronavirus via dromedary camels as the intermediate host to produce the Middle East respiratory syndrome coronavirus (MERS-CoV) in Saudi Arabia, with a mortality rate of 34% (Singhal, 2020).

3.3 WHAT HAPPENS IN THE HUMAN BODY?

Initially, the target of SARS-CoV-2 is the epithelial cells in the nose and bronchi, as well as pneumocytes. This involves the binding of the viral structural spike (S) protein and the angiotensin-converting enzyme 2 (ACE2) receptor.

The uptake of the virus is then promoted by the host type 2 transmembrane serine protease (TMPRSS2), which cleaves ACE2 to activate the S protein of the virus and hence, allow the entry of the virus.

Target cells in a host express ACE2 and TMPRSS2, especially alveolar epithelial type II cells.

In later stages, the acceleration of replication causes a breach of the epithelial-endothelial barrier integrity.

This results in the virus to infect capillary endothelial cells in the lungs to initiate the inflammatory response and cause the influx of neutrophils and monocytes.

Imaging studies from autopsies have shown thickened alveolar walls and the immune cells infiltrated in the air spaces resulting in acute respiratory distress syndrome (ARDS). The virus also attacks T lymphocyte cells leading to lymphopenia.

COVID-19 is characterized by a disrupted endothelial-epithelial barrier and damaged oxygen transmission and diffusion in the alveoli (Wiersinga et al, 2020).

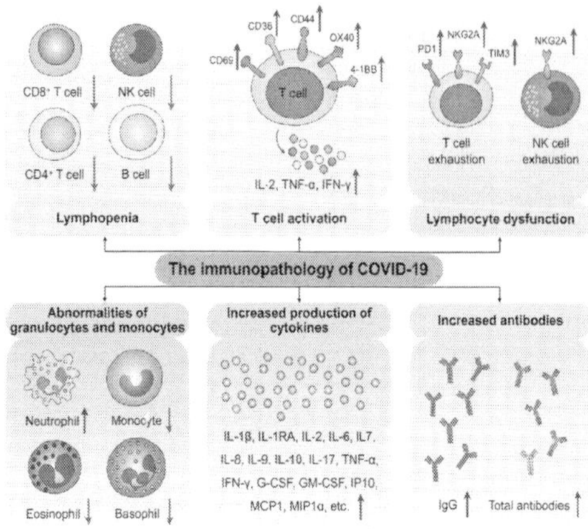

Figure 3.5. The effects of the COVID-19 virus on the immune system (*Source: Yang, L., Liu, S., Liu, J. et al.* COVID-19: *immunopathogenesis and Immuno-therapeutics. Sig Transduct Target Ther 5, 128 (2020). https://doi.org/10.1038/s41392-020-00243-2*).

An imbalance in the immune system: type 1 and type 2 T helper cells results in a cytokine storm that results in cell damage. When there is a severe infection, there is a sudden and extreme onset of clotting termed as intravascular coagulation. The formation of micro thrombi is associated with complications including pulmonary embolism, deep venous thrombosis, ischemic stroke, and myocardial infarction. Multi-organ failure can result due to sepsis (Chauhan, 2020; Wiersinga et al, 2020).

3.4 TRANSMISSION

COVID-19 is majorly transmitted by the droplets generated by coughing or sneezing or talking during face-to-face encounters. The risk is higher when exposed to symptomatic patients, such as when they cough or being within 6 feet for at least 15 minutes of an infected person. Another route is possibly through surfaces: touching a surface that has COVID-19 virus. This becomes a challenge to define given that the number of virus particles to cause an infection remains to be determined. The virus has shown to have a higher load when present on impermeable surfaces like plastic and stainless steel while the load is lower on permeable sources such as cardboard. Impermeable surfaces have shown the presence of the virus for up to 4 days post-inoculation. There is documentation of contaminated hospital rooms. The involvement of fomites (clothing or vessels or doorknobs) hence, cannot be ruled out calling for the practice of hygiene. Whether the virus is transmitted by aerosols (small droplets in the air) is yet to be established conclusively. As of now, the risk of a mother transmitting the infection to her fetus is low with most studies showing no mortalities or affected infants. The major route for transmitting the virus hence remains as face-to-face contact involving the generation of droplets. The viral load starts to reach a high 2 to 3 days before the onset of symptoms in the upper respiratory tract. There is evidence of presymptomatic transmission from infected people who were yet to "feel sick" through close contact such as group activities. There is high pharyngeal shedding in the first week of infection that causes such a rapid spread of the disease as at this time the symptoms are not really severe. Close to 48% to 62% of transmission is estimated to be caused by these presymptomatic carriers.It is yet to be clearly established as to the involvement of asymptomatic individuals given that they have mild symptoms or those who did not show symptoms when being transmitted or those who never develop symptoms at all. The nucleic acid of the virus can be detected for 6 weeks post-illness onset in throat swabs though there are reports that cultures are negative 8 days after the symptoms appear. Based on the improvement of clinical factors, an individual can be released from isolation. This isolation is for a minimum of 10 days once the symptoms appear and 3 days once these symptoms subside.

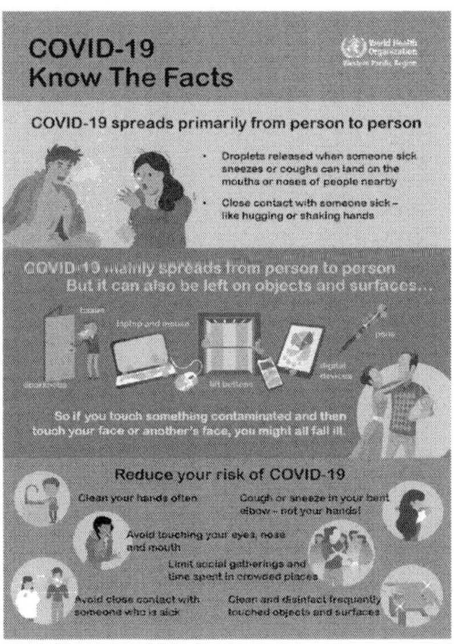

Figure 3.6. The modes of transmission of COVID-19, as stated by WHO (*Source: WHO*).

3.5 SYMPTOMS

The following are the common symptoms reported:

- 90% of the patients have fever
- 60-86% of the patients have a dry cough
- 53-80% of the patients have shortness of breath
- 15%-44% of the patients reported muscle pain
- 15%-39% had diarrhea or nausea
- 38% have fatigue or tiredness

There can also be symptoms that are not typical such as gastrointestinal symptoms. A loss of the sense of smell (olfactory dysfunction) and taste (gustatory dysfunction) has been seen in 64% to 80% of patients. Few patients (3%) show only anosmia (loss of the sense of smell) or ageusia (loss of taste) as the symptoms.

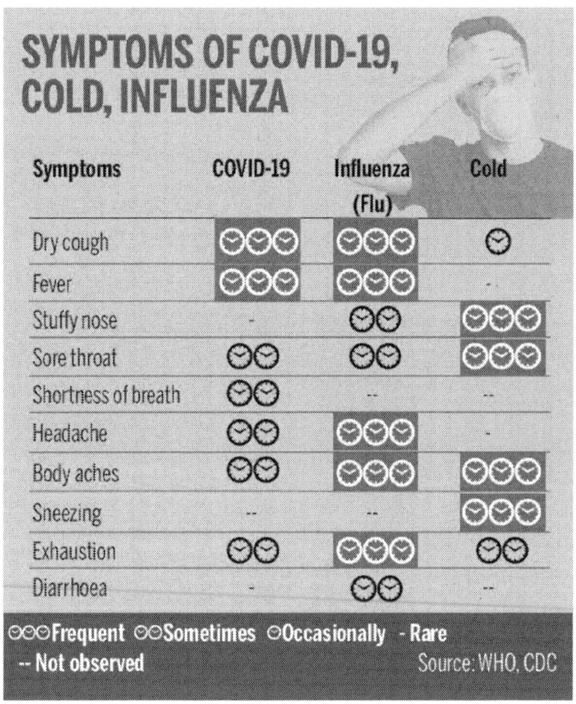
SYMPTOMS OF COVID-19, COLD, INFLUENZA

Symptoms	COVID-19	Influenza (Flu)	Cold
Dry cough	⊙⊙⊙	⊙⊙⊙	⊙
Fever	⊙⊙⊙	⊙⊙⊙	-
Stuffy nose	-	⊙⊙	⊙⊙⊙
Sore throat	⊙⊙	⊙⊙	⊙⊙⊙
Shortness of breath	⊙⊙	--	--
Headache	⊙⊙	⊙⊙⊙	-
Body aches	⊙⊙	⊙⊙⊙	⊙⊙⊙
Sneezing	--	--	⊙⊙⊙
Exhaustion	⊙⊙	⊙⊙⊙	⊙⊙
Diarrhoea	-	⊙⊙	--

⊙⊙⊙Frequent ⊙⊙Sometimes ⊙Occasionally - Rare
-- Not observed Source: WHO, CDC

Figure 3.7. The symptoms of COVID-19.

60% to 90% of patients who are hospitalized show comorbidities such as (in order of descending frequency):

- Hypertension
- Diabetes
- Cardiovascular disease
- Chronic pulmonary disease
- Chronic kidney disease
- Cancer
- Chronic liver disease

The complications associated with COVID-19 patients who are hospitalized are pneumonia in 75% of the patients; acute liver injury in 19% of patients; acute respiratory distress syndrome in 15% of patients; heart and clotting abnormalities in 10%-25%; acute kidney injury in 9% of patients; loss of consciousness in 8% of patients; shock in 6% and brain disease in 3% of patients. There are also rare complications reported such as cytokine storm and secondary hemophagocytic lymphohistiocytosis (Wiersinga et al, 2020; Zhang et al, 2020).

3.6 DIAGNOSIS

The standard approach for diagnosis is detecting the viral RNA using Reverse transcription polymerase chain reaction from the nasopharynx. However, research has shown that the timing of testing influences the sensitivity of the test. The diagnosis can also be helped by the use of serological tests such as assays and ELISA. Additionally, chest computed tomographic imaging shows the presence of peripheral ground-glass opacities in patients though many COVID positive patients show regular image results.

3.7 TREATMENT

- As of now, the support management administered to patients in the case of ADRS or acute hypoxic respiratory failure is advised. Supplemental oxygen therapy is required in 75% of the hospital-admitted patients. If there is a lack of a response to this therapy, then high-flow nasal canula oxygen can be resorted to.
- Lung-ventilation offers protection: this involves the use of a tidal volume of 4-8 mL/kg body weight; , 30 mg Hg plateau pressure. Oxygenation can also be facilitated by prone positioning or the use of muscle relaxants, such as cisatracurium.

- The use of antibacterial drugs is best recommended only for immunocompromised patients or those who have a bacterial co-infection.
- Many drugs are being explored to manage target the virus: antivirals such as remdesivir; anti-inflammatory molecules such as statins; antibodies such as plasma; anticoagulants such as heparin; immunotherapy such as tocilizumab and tyrosine kinase inhibitors. The efficacy of these molecules can depend on the stage and symptoms of the disease.
- Targeting the entry of the virus using chloroquine/ hydroxychloroquine has shown promise in vitro while exact effects are yet to be established using clinical trials.
- Other molecules being used/studied include imatinib (a tyrosine kinase inhibitor) and monoclonal antibodies against interleukin 1, interleukin 6, and interferon-gamma that are involved in mediating the inflammatory response that can damage organs.
- Another approach is the use of plasma obtained from patients who have recovered from COVID-19 (that was also used during the Spanish flu of 1918). There are reports of improvement in several parameters in ill patients such as viral load, ADRS, and body temperature.

The administration of dexamethasone (a corticosteroid) in the Randomized Evaluation of COVID-19 Therapy (RECOVERY) trial reported that 6mg of the steroid lowered the mortality in patients requiring ventilator support or with more than 7 days of symptoms (Horby et al, 2020; Chauhan, 2020; Singhal, 2020; Wiersinga et al, 2020)

- Stem cell therapy especially, the use of mesenchymal stem cells (MSCs) too has entered the arena to combat COVID-19. These cells present no ethical issues in terms of isolation, can be isolated from several tissues easily, can be expanded to clinical volumes, have shown to be safe and effective for many diseases in trials and can modulate the immune system-this point becomes important for COVID-19 given the burst of cytokines it causes (Golchin et al, 2020).

To see the potential of MSCs currently, the following studies have been presented: 7 enrolled patients with COVID-19 pneumonia in Beijing YouAn Hospital, China administered MSCs showed an improvement in pulmonary function and symptoms in 2 days of transplantation with no adverse effects.

The MSCs lowered the overactivated immune system seen in COVID-19 making the conditions favorable for healing (Zhao et al, Aging and Disease, 2020). A critical female patient with COVID-19 who received umbilical cord MSCs along with antibiotics and α1 thymosin showed relief in symptoms and was removed from the ICU ward showing the healing effect of MSCs for treatment (Liang et al, chinaXiv, 2020).

3.8 PREVENTION

COVID-19 can be prevented by following hygiene and social distancing. Countries across the globe have followed and implemented several systems to check and prevent the incidence and spread of the virus. These approaches include:

- Personal hygiene, hand washing, the use of face masks/shields, and physical distancing. The spread of the infection is lowered by the use of face masks. The use of greater distancing between people offers better protection from the disease. Proper disinfection of the self and the environment is also vital.
- Regulation such as stay at home or work from home, school closure, limiting the operations of businesses or shops, or public gatherings. Closed public transport is another point.
- Efficient identification and tracking of cases
- Closing borders and enforcing quarantine

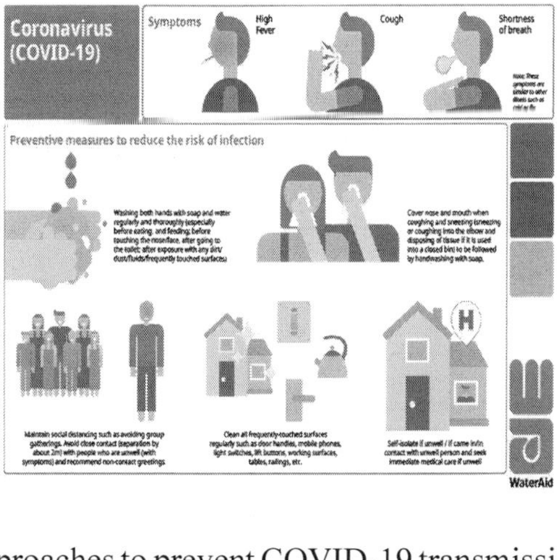

Figure 3.8. The approaches to prevent COVID-19 transmission (*Source: https://www.wateraidindia.in/covid19*).

Tests are ongoing for a SARS-CoV-2 vaccine in humans with no vaccine in hand as of yet. These approaches range from targeting the S protein to the nucleic acid to recombinant systems. Points that require further research include immunogenicity, design of the domain used, dosage, safety, efficacy, large-scale production, and legal aspects such as technology transfer and licenses.

3.8.1 Author's note

An important factor here is to control the infection while balancing the damage to the economy and society. The optimal approach would depend on literacy, government policies, culture, population, and resources. There are several questions that still need answers backed by solid research such as:

- Immunity to the virus
- The influence of the diet or culture on immunity
- The efficacy and safety of treatment
- Herbs as medicine
- The potential of multiple therapies such as conventional and traditional medicines such as Ayurveda

Together, we overcome is the motto we are looking at!

3.9 REFERENCES

1. Chauhan S (2020) Comprehensive Review of Coronavirus Disease 2019 (COVID-19). Biomed J. https:// doi.org/10.1016/j. bj.2020.05.023, ISSN 2319-4170
2. Wiersinga WJ, Rhodes A, Cheng AC, Peacock SJ, Prescott HC (2020) Pathophysiology, Transmission, Diagnosis, and Treatment of Coronavirus Disease 2019 (COVID-19): A Review. JAMA. 10.1001/jama.2020.12839. doi:10.1001/jama.2020.12839.
3. Singhal T. (2020) A Review of Coronavirus Disease-2019 (COVID-19). Indian J Pediatr. 87:281–286. https://doi.org/10.1007/s12098-020- 03263-6.
4. Zimmermann P, Curtis N (2020) Coronavirus Infections in Children Including COVID-19. Pediatr Infect Dis: 39(5):355-368. doi: 10.1097/ INF.0000000000002660.
5. Zhang Q, Shan KS, Abdollahi S, et al. (May 01, 2020) Anosmia and Ageusia as the Only Indicators of Coronavirus Disease 2019 (COVID-19). Cureus 12(5): e7918. doi:10.7759/cureus.7918.
6. Horby P, Lim WS, Emberson J, Mafham M, Bell J, et al. (2020) Effect of Dexamethasone in Hospitalized Patients with COVID-19: Preliminary Report. medrxiv. Published online June 22, 2020. doi:10.1101/2020.0 6.22.20137273:24
7. Golchin A, Seyedjafari E, Ardeshirylajimi A (2020) Mesenchymal Stem Cell Therapy for COVID-19: Present or Future. Stem Cell Rev. 1–7.
8. Leng Z, Zhu R, Hou W, Feng Y, Yang Y, Han Q, et al. (2020) Transplantation of ACE2- Mesenchymal Stem Cells Improves the Outcome of Patients With COVID-19 Pneumonia. Aging Dis. 11(2):216. 10.14336/ad.2020.0228.
9. Liang B, Chen J, Li T, Wu H, Yang W, Li Y, et al. (2020) Clinical Remission of a Critically Ill COVID-19 Patient Treated by Human Umbilical Cord Mesenchymal Stem Cells. Medicine (Baltimore). 99(31):e21429. Doi:10.1097/MD.0000000000021429.

4

Zika virus: From Uganda to Brazil

Shivsanjeevi Sripathi

ShivSanjeevi Sripathi completed his Masters in Biotechnology from Mumbai University in 2008. He was awarded for academic excellence in both his Bachelors and Masters for securing second rank in Mumbai University in 2006 and first rank in his college: Kishinchand chellaram College. For his Masters he secured first rank in his college KET's V.G.Vaze College. He qualified CSIR and NET and and TOEFL in September 2008. He then worked on a stem cell project at the Specialized Centre for Cell Based Therapy (SCCT),KEM Hospital at Mumbai on a project entitled, : Isolation & detection of stem cells from Human Umbilical cord/ amniotic membrane" following which he worked at Junior Research Fellow at Microbiology & Cell Biology Department, Indian Institute of Science, Bangalore from August 2009 to November 2011. The field of work involved cloning of cell wall genes and transcription factors in *E.coli* & *M.smegmatis*. As a writer, he has authored and co-authored 25 books with Delve and Arcler International Press. He loves to read and share on interesting aspects of life sciences in books. In his free time he loves to travel and explore ancient customs and traditions.

4.1 BASICS

Zika virus, of the *Flaviviridae* family of viruses, is an emergent virus borne by mosquitoes. This family includes the following genera Flavivirus, Hepacivirus, Pegivirus, and Pestivirus of which, Zika belongs to the first one. This genus also has other viruses that cause a range of diseases such as Yellow fever, encephalitis (Murray Valley, Japanese, tick), and dengue. Such viruses cause either mild to life-threatening diseases. These viruses have a single-stranded RNA strand as genetic material positive covered by an envelope.

A rhesus macaque from the Zika forest in Uganda was the first source from which the Zika virus was isolated. This was followed by detecting the presence of neutralizing antibodies against the virus in human sera samples in Uganda. This showed the ability of the virus to infect humans too. While initially it was thought that the Zika virus was limited only in Africa and Asia, infections were also reported in the Pacific islands and South America. Further research was fueled by the possibility of the virus to affect the nervous system. In 2013, French Polynesia saw an outbreak of the virus associated with the Guillain– Barré syndrome.

Figure 4.1. Geographical indication of Zika virus spread (*Source: https://healthywa.wa.gov.au/Articles/U_Z/Zika-virus)*.

There was a sudden burst of infections in Brazil in 2015 that was initially presented as a "dengue-like syndrome" that was confirmed to be Zika virus using reverse transcription polymerase chain reaction (RT-PCR) assay and DNA sequencing. This spread to Central, North and South America. Given that these infections also manifested as Guillain–Barré syndrome and microcephaly, the WHO declared this outbreak as a Public

Health Emergency of International Concern (PHEIC). The Zika virus strain of Brazil and the one of French Polynesia shared a common ancestor (Noorbakhsh et al, 2019).

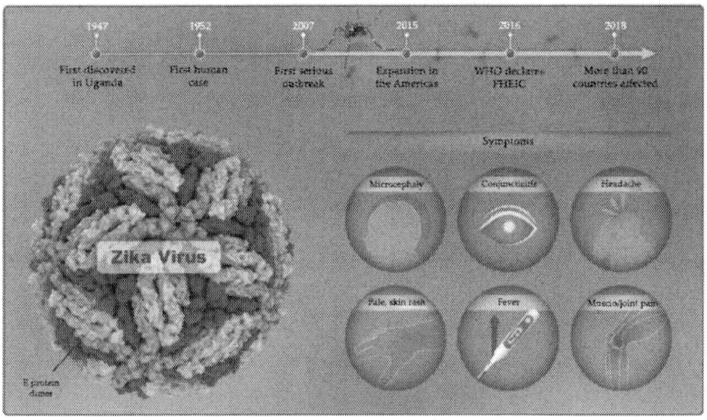

Figure 4.2. The overview of the virus and its spread (Noorbakhsh et al, 2019).

4.2 HOW DOES THE DISEASE SPREAD AND STRIKE?

The pathogenesis of the Zika virus is on the same lines as other Flavivirus infections. It is an arbovirus that is transmitted between vertebrates by a mosquito bite. When an Aedes mosquito bites an infected person, it ingests the Zika virus that it spreads to another human. When the mosquito ingests the Zika virus, the latter begins its replication in the midgut epithelial cells from where it spreads to the salivary glands of the mosquito. 10 days after this, the insect becomes a vector capable of infecting a human.

Research has shown that apart from humans, apes, monkeys, elephants, goats, and sheep too show the presence of antibodies. Once the virus is transmitted via the mosquito bite, many cells can be infected such as dendritic cells, dermal fibroblasts, and skin keratinocytes. This infection is thought to be mediated by receptors such as Tyro3, AXL, and DC-SIGN. The completion of replication is followed by the spread of the virus to lymph nodes while infecting the nervous system too. The virus can also penetrate the placental blood barrier to infect a fetus in pregnant women.

The entry of the Zika virus in a cell has shown to induce interferon's in the cells: mostly interferon-beta. The innate system is also activated: mainly nucleic acid-sensing "pattern recognition receptors" (PRRs). Fibroblasts infected with the Zika virus show high levels of RIG-I and MDA-5: molecules that get the signaling pathway for viral RNA molecule detection started off. This is followed by adaptive immune responses where the migration of dendritic cells infected with the virus to the lymph nodes activates T-cells.

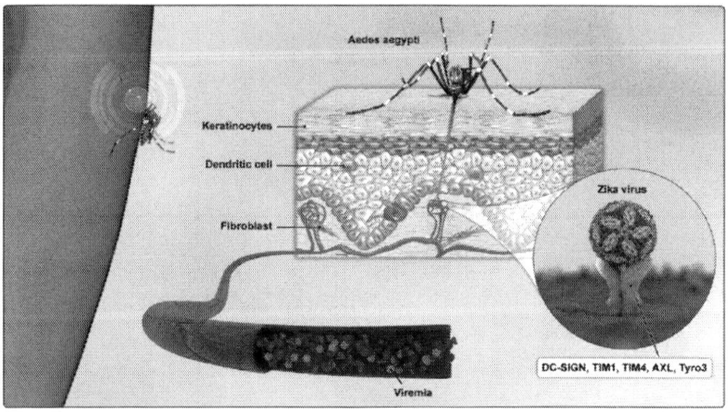

Figure 4.3. The pathogenesis of the Zika virus (Noorbakhsh et al, 2019).

The Zika virus can infect monocytes in the fetus to, in turn, target the developing nervous system. The neural cells of aborted fetuses have shown the presence of the virus. In adults, the virus is thought to directly target the nervous system cells and also indirectly by causing the activation of microglial cells and infiltration of macrophages. The virus is also implicated to be a cause of a CNS disease called "Acute Disseminated Encephalomyelitis" (ADEM).

The symptoms of Zika virus infection are on the lines of that of chikungunya and dengue. Post 3-12 day incubation, there is fever, headache, red eyes, and a maculopapular rash (Agumadu and Ramphul, 2018; Noorbakhsh et al, 2019).

4.3 DIAGNOSING THE DISEASE

4.3 1 Serological Assays

There is the development of an IgM antibody against the Zika virus generally in 7 days of infection that can be quantified from Day 5 onwards. This is followed by an anti-Zika virus IgG antibody whose detection time varies between months to years' post-infection.

Tests include hemagglutination inhibition, ELISA, complement fixation, neutralization, and immunofluorescence.

The following assays have received approval from the FDA to detect the IgM antibody for emergency use:

- IgM antibody capture ELISA (Zika MAC-ELISA) (CDC) ,
- ZIKV Detect IgM capture ELISA (InBios International, Inc.),
- Liaison XL Zika capture IgM assay (DiaSorin Incorporated),
- ADVIA Centaur Zika test (Siemens Healthcare Diagnostics Inc.)
- DPP Zika IgM system (Chembio Diagnostic Systems, Inc.)

An important point to be notes here is that as the Zika virus test can show cross-reactivity with other viruses of the same family (dengue), the above tests require careful analyses and confirmation using by plaque-reduction neutralization testing (PRNT).

4.3.2 Molecular Assays

The RNA of Zika virus can be detected in the serum, plasma, saliva, urine, breast milk, semen, amniotic fluid, conjunctival fluid, as well as the fetus tissues.

The gold standard for diagnosis is reverse transcriptase PCR (RT-PCR). The Cobas Zika test (Roche) has received FDA approval for use in testing. In the case of an emergency, the following assays have been approved by the FDA:

- Triplex Real-Time RT-PCR that detects Zika virus, Dengue virus, and Chikungunya virus (CDC),
- Zika virus RNA qualitative real-time RT-PCR (Quest Diagnostics Infectious Disease, Inc.),
- RealStar Zika virus RT-PCR kit (Altona Diagnostics, GmbH)

4.4 TREATMENT

Currently, an antiviral treatment for the Zika virus infection is still being sought after. Several approaches are being tested ranging from looking at earlier-used drugs or *in silico* designed drugs (Agumadu and Ramphul, 2018; Noorbakhsh et al, 2019).

- An anti-bacterial molecule nanchangmycin and *in silico* designed ZINC33683341 and ZINC49605556 are aimed to block the attachment of the virus to target cells.
- The cationic squalamine targets the electrostatic interactions between the target cell and the virus.

- The anti-malarial drug chloroquine targets endolysosomal vesicles to finally block endosomal fusion.
- Berberine that binds the dengue virus protease also targets the Zika virus NS3 protease (a helicase that functions in viral replication).
- An FDA approved nucleotide analog inhibitor called Sofosbuvir used for HCV also targets Zika virus replication.

4.5 TESTING FOR THE VIRUS

As per the recommendations of the Centers for Disease Control and Prevention (CDC), testing of the following people is vital (Agumadu and Ramphul, 2018):

- People with recent exposure or symptoms
- Pregnant women who have been or are being exposed to Zika virus
- Serum Zika virus and dengue virus immunoglobulin M (IgM) test and serum and urine testing for Zika virus ribonucleic acid (RNA) test and nucleic acid testing (NAT) is to be done for a non-pregnant patient showing symptoms. NAT is best advised within 2 weeks of showing the symptoms.
- NAT in serum and urine and IgM in the serum should be tested in a pregnant woman showing symptoms. The test is positive if the IgM test is negative but the NAT is positive. If only the IgM test is positive then the plaque reduction neutralization test (PRNT) is required for confirmation.
- A pregnant woman with no symptoms but exposed currently to the Zika virus requires NAT testing in the first visit followed by two non-consecutive tests.
- The testing in the case of a pregnant woman with no symptoms with recent exposure to the virus should be on the basis of the case presented.
- Ultrasound diagnosis of congenital Zika virus in a pregnant woman should be tested as described for pregnant women showing symptoms. The testing of the fetal tissues and placenta can be looked at based on the scenario.
- The testing has not been advised for asymptomatic individuals.

4.6 PREVENTION AND CONTROL OF THE SPREAD OF THE VIRUS

As the virus is spread by the *Aedes* mosquitoes, the first step is to target these insects. As these mosquitoes are active both in the morning and night, the use of safe mosquito repellents is warranted. These repellants should be safe on the elderly and pregnant or breastfeeding women. It is not recommended to apply these repellants to infants lower than 2 months in age. Similarly, para-menthane-diol (PMD) or oil of lemon eucalyptus is not recommended for children below 3 years of age.

Apart from repellants, the use of long clothes to cover hands and legs is an effective option. The use of mosquito nettings correctly is another option. Avoiding the presence of stagnant water in and around the house is vital.

The prevention of diseases borne by *Aedes* requires careful collaboration between the academic and health authorities along with the community.

The presence of the virus in the vaginal fluid and semen of infected people makes the following of safe practices or abstinence advisable. A pregnant woman should avoid visiting high-risk areas.

Table 4.1. Areas and countries potentially at risk of Zika (Noorbakhsh et al, 2019):

Africa	Angola, Benin, Burkina-Faso, Burundi, Cameroon, Cape Verde, Central African Republic, Chad, Congo (Congo-Brazzaville), Côte d'Ivoire, the Democratic Republic of the Congo (Congo-Kinshasa), Equatorial Guinea, Gabon, Gambia, Ghana, Guinea, Guinea-Bissau, Kenya, Liberia, Mali, Niger, Nigeria, Rwanda, Senegal, Sierra Leone, South Sudan, Sudan, Tanzania, Togo, Uganda
Asia	Bangladesh, Burma (Myanmar), Cambodia, India, Indonesia, Laos, Malaysia, Maldives, Pakistan, Philippines, Singapore, Thailand, Timor-Leste (East Timor), Vietnam
The Caribbean	Anguilla; Antigua and Barbuda; Aruba; Barbados; Bonaire; British Virgin Islands; Cuba; Curaçao; Dominica; Dominican Republic; Grenada; Haiti; Jamaica; Montserrat; the Commonwealth of Puerto Rico, a US territory; Saba; Saint Kitts and Nevis; Saint Lucia; Saint Martin; Saint Vincent and the Grenadines; Sint Eustatius; Sint Maarten; Trinidad and Tobago; Turks and Caicos Islands; US Virgin Islands
Central America	Belize, Costa Rica, El Salvador, Guatemala, Honduras, Nicaragua, Panama
North America	Mexico

The Pacific Islands	Fiji, Papua New Guinea, Samoa, Solomon Islands, Tonga
South America	Argentina, Bolivia, Brazil, Colombia, Ecuador, French Guiana, Guyana, Paraguay, Peru, Suriname, Venezuela

Care for infected patients should involve minimum physical contact and proper hand washing.

An ethical concern that enters the picture is the possibility of infection in pregnant women as mothers with Zika virus infection in their first trimester of pregnancy have delivered children with congenital issues (microcephaly). The laws regarding the abortion of such fetuses are different across nations (Agumadu and Ramphul, 2018; Noorbakhsh et al, 2019).

4.7 FINAL POINTS

Prophylaxis or vaccines against the Zika virus infection are still under research. Thus, the correct education of people especially, in underdeveloped or developing countries is vital. Early detection and correct management of the disease can help handle epidemics (Agumadu and Ramphul, 2018; Noorbakhsh et al, 2019).

4.8 REFERENCES

1. Agumadu V C, Ramphul K (July 22, 2018) Zika Virus: A Review of Literature. Cureus 10(7): e3025. DOI 10.7759/cureus.3025.
2. Noorbakhsh, F., Abdolmohammadi, K., Fatahi, Y., Dalili, H., Rasoolinejad, M., Rezaei, F., Salehi-Vaziri, M., Shafiei-Jandaghi, N. Z., Gooshki, E. S., Zaim, M., & Nicknam, M. H. (2019). Zika Virus Infection, Basic and Clinical Aspects: A Review Article. Iranian journal of public health, 48(1), 20–31.

5

Dengue viruses: From the Tropics to a Global Health Challenge

Shivsanjeevi Sripathi

ShivSanjeevi Sripathi completed his Masters in Biotechnology from Mumbai University in 2008. He was awarded for academic excellence in both his Bachelors and Masters for securing second rank in Mumbai University in 2006 and first rank in his college: Kishinchand chellaram College. For his Masters he secured first rank in his college KET's V.G.Vaze College. He qualified CSIR and NET and and TOEFL in September 2008. He then worked on a stem cell project at the Specialized Centre for Cell Based Therapy (SCCT),KEM Hospital at Mumbai on a project entitled, : Isolation & detection of stem cells from Human Umbilical cord/ amniotic membrane" following which he worked at Junior Research Fellow at Microbiology & Cell Biology Department, Indian Institute of Science, Bangalore from August 2009 to November 2011. The field of work involved cloning of cell wall genes and transcription factors in *E.coli & M.smegmatis*. As a writer, he has authored and co-authored 25 books with Delve and Arcler International Press. He loves to read and share on interesting aspects of life sciences in books. In his free time he loves to travel and explore ancient customs and traditions.

5.1 DENGUE

Belonging to the family Flaviviridae and the genus *Flavivirus*, the dengue virus is vector-borne (arthropod) to be considered as a public health challenge globally in the tropics and sub-tropics. There has been a 30-fold increase between 1960 and 2010 in dengue; this is attributed to global warming, the growth in population, poor planning of cities, excessive travel, and inefficient mosquito control. 2.5 billion people have made dengue-endemic regions their homes. The first dengue case has been traced back to 1780 in Madras (now Chennai) in India where a similar illness was reported. The years 1963-1964 saw the first proven epidemic in Calcutta and the Eastern Coast (Guzman et al, 2010; Hasan et al, 2016).

While the disease has become endemic in several countries such as the Americas, Africa, the Western Pacific, South-East Asia, and the Eastern Mediterranean, dengue was endemic in only 9 countries before 1970. Asia bears 70% of the world's disease burden. In the case of dengue, the disease is entering new areas causing a spike in the number of cases. The year 2010 saw the first report of local transmission in France and Croatia. More than 2000 cases were seen in an outbreak in 2012 in the Madeira islands of Portugal. Dengue is ranked as the second cause of fever (the first being malaria) in travelers from low/middle-income countries. Even European countries are reporting autochthonous cases every year (WHO data).

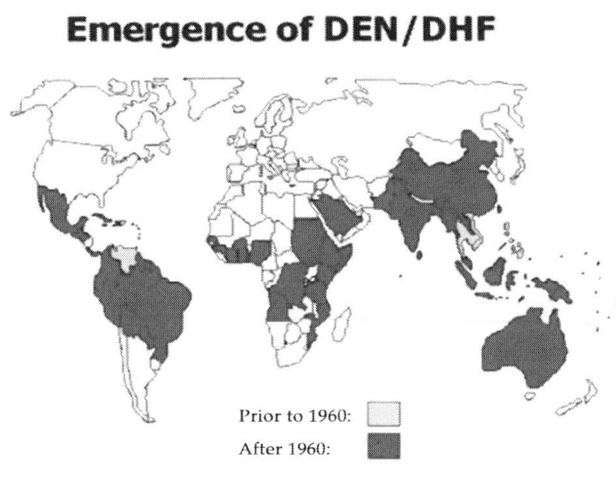

Figure 5.1. The global spread of dengue over time (*Source: https://www.who.int/csr/disease/dengue/dehngueemergence.jpg*).

Over 2.38 million cases were reported in the Region of the Americas in 2016 with 1.5 million cases in Brazil. The burden on the Philippines and Malaysia was 176 411 and 100 028 cases, respectively. There was a drop in the number of cases in the 2017 post-Zika period, in these countries.

The year 2019 saw the largest number of cases of dengue globally with even Afghanistan entering the scenario of disease. 3.1 million cases were reported in the American region. 420,000 cases were reported in the Philippines while Bangladesh reported 101,000 cases and Malaysia reported 131,000. However, the number of mortalities was lesser than in previous years.

Dengue remains an issue in 2020 too with increases in Bangladesh, Brazil, Cook Islands, Ecuador, India, Indonesia, Maldives, Mauritania, Mayotte (Fr), Nepal, Singapore, Sri Lanka, Sudan, Thailand, Timor-Leste, and Yemen (WHO Fact Sheet).

5.2 THE VIRUS

Among the 50-100 million global cases of dengue fever (DF), half a million are what is called dengue hemorrhagic fever (DHF). This is seen mainly in children below 15 years of age with fever and abnormally low levels of platelets or thrombocytopenia.

The virus is enveloped, icosahedral with a single-stranded, positive-sense RNA. The C protein around the genome is located in the icosahedral core that is forty to fifty nanometers in diameter. The core is surrounded by the viral proteins, M and E.

There are 4 serotypes of the dengue virus (DENV): DENV-1, DENV-2, DENV-3, and DENV-4-there is no immunity among the serotypes i.e. infection by one does not confer immunity to another serotype. The virus is transmitted by the *Aedes* mosquitoes; the major one being *Aedes aegypti* followed by *Aedes albopictus* (Hasan et al, 2016; Sanyaolu et al, 2017).

5.3 WHAT HAPPENS DURING THE BITE AND DISEASE?

The dengue virus (DENV) is injected into a human host during the blood meal by a mosquito. Receptor-mediated endocytosis of the virus results in immature Langerhans cells and keratinocytes getting infected by the virus. The viral replication is controlled by non-structural proteins, NS1, and NS2A with the former also activating the complement system cascade. These infected cells can migrate to lymph nodes to cause the recruitment of macrophages or undergo apoptosis.

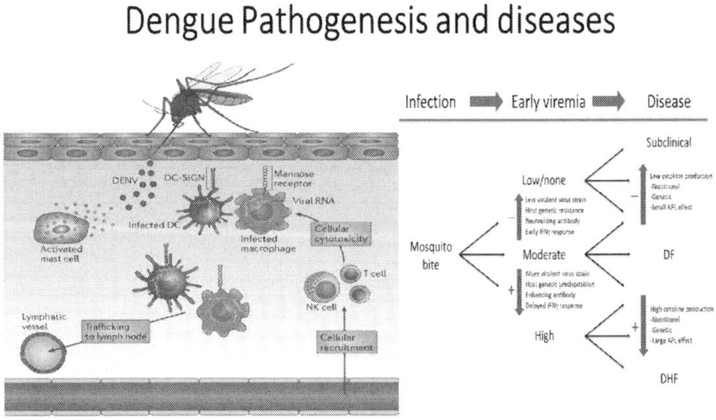

Figure 5.2. The pathogenesis of the dengue infection (*Source: https://www.onlinebiologynotes.com/dengue-pathogenesis-clinical-manifestation-lab-diagnosis-treatment/*).

5.4 SYMPTOMS

The infection of DENV-1 to DENV-4 can be quelled by the immune system resulting in asymptomatic conditions. However, there is a broad range of symptoms also possible.

5.4.1 Undifferentiated fever

This is observed in primary infections and sometimes in an initial secondary infection. It is challenging to distinguish it from other types of viral fevers.

5.4.2 Dengue fever (DF)

Both primary and secondary infections show DF where there is a high-grade fever lasting for 3 days to 1 week. There is also a severe headache, muscle pain, loss of appetite, painful joints, stomach pain, metallic taste in the mouth, and tiredness. The muscle pain has earned dengue the name break bone fever. 50-80% of the patients show a unique skin rash: facial flushing caused by capillary dilatation in the first 2 days of fever. There is a second rash seen 3 days to 1 week following the fever as eruptions often described as "white islands in a sea of red." There are instances of epistaxis (bleeding from the nose) and gingival bleeding (from the gums).

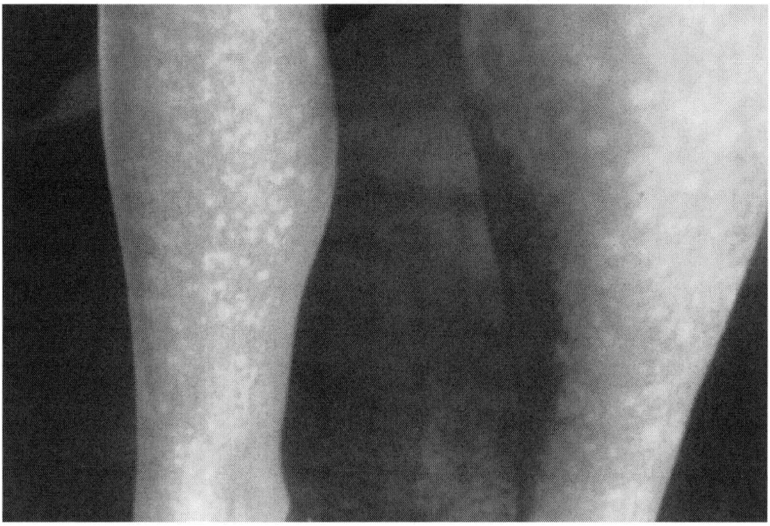

Figure 5.3. The skin rash seen in dengue fever (*Source: https://www.emedicine-health.com/dengue_fever/article_em.htm).*

5.4.3 Dengue Hemorrhagic fever (DHF)

This can be of the following types:

Grade I: Only mild bruising or a positive tourniquet test

Grade II: Spontaneous bleeding into the skin and elsewhere

Grade III: Clinical signs of shock

Grade IV: Severe shock

Clinically, DHF is a continuous fever of 2 to 7 days with skin or mucosal bleeding and thrombocytopenia (<100,000 cell/mm^3). It is often seen in secondary dengue infections or in infants as a primary infection.

In the Acute-onset febrile phase, there is a high-grade fever (2-7 days) with bleeding from the nose or gums. There is a positive result for the tourniquet test and hepatomegaly. The destruction of platelets and lowered production of platelets results in thrombocytopenia. This makes the blood vessels fragile to cause the hemorrhage observed.

The major signs of DHF are the bleeding observed directly or through the positive tourniquet test and plasma leakage from pleural or peritoneal surfaces.

DHF is also associated with facial features such as crusting of lips and tongue and erythema. Eruptions on the soft palate or hemorrhagic bullae on the tongue and floor of the mouth have also been reported. The buccal mucosa can show the presence of brown lesions that can bleed on touch.

5.4.4. Dengue shock syndrome

When DHF is seen with an unstable pulse, cold and clammy skin, cyanosis and restlessness, it is known as shock syndrome that manifests later as shock and multi-organ damage.

The infection of a host with a second serotype causes a secondary infection that triggers the immune system to cause this severe form of manifestations. The antibody-dependent enhancement (ADE) theory attributes this to the binding of antibodies generated from a first infection due to cross-reactivity to bind the new serotype. Instead of neutralizing the new serotype, the virus replicated to activate the production of cytokines that cause the permeability of blood vessels and associated bleeding.

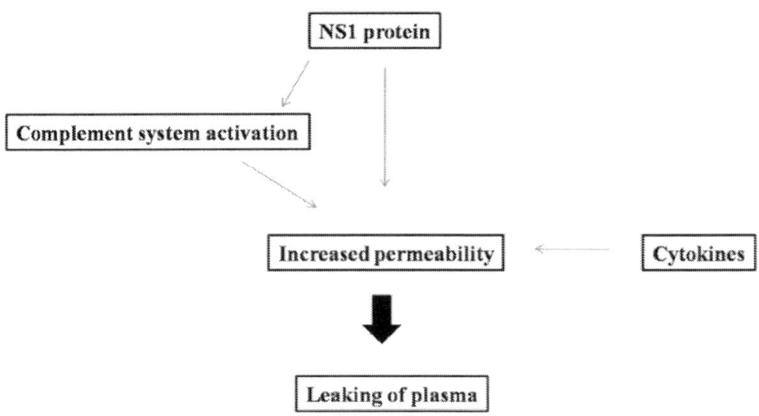

Figure 5.4. The immune system response and the effects on increased vascular permeability and leakage.

Mast cells release anti-viral molecules and also recruit other cells of the immune system to kill the dengue virus. Sometimes, the virus can reach other organs via the blood or lymph to activate the mast cells in the spleen or lymph nodes. These activated cells release inflammatory molecules that dilate the capillaries and cause plasma leakage.

The disease severity is dependent on the types of cytokines released mainly: Interferon γ (IFN-γ) that upregulates Fc receptors on target cells to cause the virus uptake and hence, observed symptoms. Other molecules such as tumor necrosis factor α (TNF- α), interleukin 2 (IL-2), interleukin-8 (IL-8), etc. have also been implicated.

5.5 DIAGNOSIS AND TREATMENT

A key factor for efficient recovery is the early and correct diagnosis. Diagnosis can be done for viral RNA, serology, or detecting the dengue virus by culture. Testing for the NS1 antigen and antibody and real time reverse transcription-polymerase chain reaction (RT–PCR) are used to check for the presence of the virus. Using viral antigens for immunohistochemistry and immunofluorescence can confirm the presence of the virus.

A specific treatment regime for dengue is yet to be established. As the hallmark of this disease is increased vascular permeability and associated plasma loss, maintaining the fluid and electrolyte balance in a patient is the major approach. This involves the intravenous use of whole blood or Ringer's lactate or Ringer's acetate or saline or plasma or plasma substitutes or 5% glucose solution.

Platelet transfusions become important when there is significant hemorrhage: <10,000/cu mm platelets.

Oral solutions can help alleviate thirst. The use of aspirin is not recommended however, acetaminophen can be used for addressing the fever.

A patient is discharged from a medical center if he/she shows a platelet count >50,000/mm^{3,} optimal appetite, no fever, and adequate urine production.

5.6 PREVENTION

Targeting *Aedes aegypti*, the vector for the dengue virus is the primary course of action. This involves regular fumigation and removing abandoned containers that are foci for breeding. People who travel to or live in dengue-endemic areas should avoid the presence of stagnant water in and around the houses, use repellants, cover themselves properly when stepping out, and correctly use mosquito nets and screens.

The major factors of urbanization and associated travel are responsible for the spread of the virus. Additionally, the genetic variation in the virus contributes to its increased virulence (Hasan et al, 2016; Sanyaolu et al, 2017).

As far as vaccines are concerned, the infection is augmented by antibodies to the pre-membrane (prM) protein and the fusion loop (FL) epitope of the envelope (E) protein. Additionally, the efforts and research to develop a dengue vaccine are focused on the antibody-dependent enhancement (ADE). There is the chimeric yellow fever/dengue tetravalent dengue vaccine (CYD-TDV): a live attenuated vaccine developed by Sanofi and is marketed as Dengvaxia. However, the vaccine has been recommended only for seropositive individuals by the WHO Global Advisory Committee on Vaccine Safety.

Other vaccines under development currently include:

- TDENV-PIV by WRAIR and GSK is in Phase I/II trials (Inactivated vaccine)
- TetraVax-DV by Butantan is in Phase III trial (Live attenuated vaccine)
- TDV by Takeda is in Phase III trial (Live attenuated vaccine)
- Dengue tetravalent vaccine by Panacea Biotec Ltd is in Phase I/II trials (Swaminathan and Khanna, 2019)

5.7 FINAL POINTS

From an initially endemic disease, dengue has reached places where it was previously not present. While the global estimate for dengue incidence stands at 50 to 200 million, cartography places the numbers at 400 million! The disease is considered to inflict more human suffering and is the number one cause of hospitalization and death in Southeast Asian children. One should also take into account the financial burden associated with the disease. The effect of temperature changes and globalization is expected only to cause an increase in the number of cases. Care should be taken by healthcare professionals while diagnosing and treating this disease.

5.8 REFERENCES

1. Hasan S, Jamdar SF, Alalowi M, Al Ageel Al Beaiji SM (2016) Dengue virus: A Global Human Threat: Review of Literature. Journal of International Society of Preventive & Community Dentistry, 6(1), 1–6. https://doi.org/10.4103/2231-0762.175416
2. Guzman MG, Halstead SB, Artsob H, Buchy P, Farrar J, Gubler DJ, et al. (2010) Dengue: A continuing global threat. Nat Rev Microbiol. 8(Suppl): S7–16.
3. World Health Organization. Dengue and severe dengue. Available on: https://www.who.int/news-room/fact-sheets/detail/dengue-andsevere-dengue
4. Sanyaolu A, Okorie C, Badaru O, Adetona K, Ahmed M et al. (2017) Global Epidemiology of Dengue Hemorrhagic Fever: An Update. Review Article 5(6): 00179. DOI: 10.15406/jhvrv.2017.05.00179
5. Swaminathan S, Khanna N (2019) Dengue Vaccine Development: Global and Indian Scenarios. International Journal of Infectious Diseases 84S (2019) S80–S86.

6

Ebola: From an Exotic Virus to an Epidemic

Shivsanjeevi Sripathi

ShivSanjeevi Sripathi completed his Masters in Biotechnology from Mumbai University in 2008. He was awarded for academic excellence in both his Bachelors and Masters for securing second rank in Mumbai University in 2006 and first rank in his college: Kishinchand chellaram College. For his Masters he secured first rank in his college KET's V.G.Vaze College. He qualified CSIR and NET and and TOEFL in September 2008. He then worked on a stem cell project at the Specialized Centre for Cell Based Therapy (SCCT),KEM Hospital at Mumbai on a project entitled, : Isolation & detection of stem cells from Human Umbilical cord/ amniotic membrane" following which he worked at Junior Research Fellow at Microbiology & Cell Biology Department, Indian Institute of Science, Bangalore from August 2009 to November 2011. The field of work involved cloning of cell wall genes and transcription factors in *E.coli & M.smegmatis*. As a writer, he has authored and co-authored 25 books with Delve and Arcler International Press. He loves to read and share on interesting aspects of life sciences in books. In his free time he loves to travel and explore ancient customs and traditions.

6.1 THE VIRUS AND ITS EPIDEMICS

Belonging to the family Filoviridae, the members of the genus *Ebolavirus* are enveloped, single-stranded, enveloped RNA viruses. This genus along with another *Marburgvirus* genus of the same family is the cause of similar diseases. As they cause symptoms of clotting defects, capillary leakage, and shock, they are hemorrhagic fever viruses.

The *Ebolavirus* genus includes the following that shows differences in virulence:

Tai Forest Ebolavirus has not produced any significant infection or fatality

Reston Ebolavirus (Reston virus, RESTV), discovered in Reston, Virginia in macaques from the Philippines. These viruses cause disease in primates and no symptomatic disease in humans.

Sudan Ebolavirus (SUDV) causes mortality of 41–65%.

Bundibugyo ebolavirus (Bundibugyo virus) causes mortality of 40%.

Zaire ebolavirus (Ebola virus, EBOV) was detected in Zaire or the Democratic Republic of Congo, Africa in 1976. As it was first observed near the Ebola River in Congo, it got the name Ebola.

There was another occurrence in 1979 in Sudan. The majority of the outbreaks were noticed in the Democratic Republic of the Congo, Gabon, and the Republic of the Congo till 2013.

The largest epidemic as of today was seen between late 2013 and early 2016. Till now, the virus was considered as exotic, but the loss of 11,325 people and 28,652 human infections transformed its status to a pathogen that can affect global health.

The WHO declared a Public Health Emergency of International Concern and the epidemic was brought under control. Ebola (virus disease) Treatment Units (ETUs) were established where scientists and doctors were involved to control patients.

The damage to families and the economy was immense. The transmission was widespread: 28,616 cases with 11,310 deaths in Liberia, Sierra Leone, and Guinea. Cases due to travel were reported in Italy, Mali, Nigeria, Senegal, Spain, the United Kingdom, and the United States of America: 36 cases with 15 deaths (Kourtis et al, 2015; Singh et al, 2017; Hasan et al, 2019; Ebola Primer, 2020).

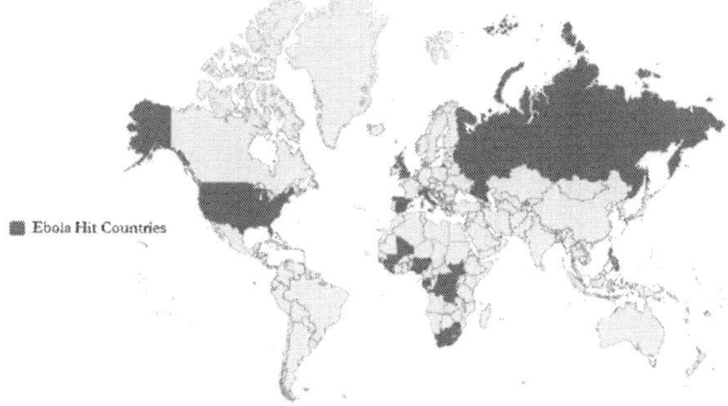

Figure 6.1. The spread of the Ebola virus disease between 1976-2016 (Singh et al, 2017).

Apart from this epidemic, an epidemic has been reported in 2020 in the Democratic Republic of the Congo: with 3463 cases and 2280 deaths (WHO data: 9th June 2020). The Ebola virus is a "biosafety level 4 and category A bioterrorism pathogen" capable of nationwide spread, according to the Centers for Disease Control and Prevention (CDC).

6.2 THE SPREAD

In Africa, the natural hosts are *Hypsignathus monstrous, Epomops franqueti*, and *Myonycteris torquata*: fruit bats from the *Pteropodidae* family. When the fruits partially eaten by bats are consumed by non-human primates, the disease is transmitted among themselves and also to humans.

Figure 6.2. The schematic of transmission of the Ebola virus.

The transmission of the virus involves close bodily contact with an infected person or contact with their clothing (alive or dead) or fluids. One of the major causes of the 2014-outbreak in Guinea was the funeral ceremonies in the African countries. These involve direct handling of the deceased. Close to 68% of the infections in 2014 were due to this activity (Chan, 2014).

While transmission through the air is not suggested and yet is not clear, intimate contact with patients and their body fluids is the major cause of the spread. There are also pieces of evidence of sexually transmitting the virus.

The RNA has been detected from feces, tears, rectal swabs, vagina, skin with the vaginal, rectal, and conjunctival samples showing its presence for a month. The virus has also been detected in semen for up to 3 months, suggesting its presence in patients who are recuperating.

Filipino pigs have shown the presence of the virus and the habit of consuming uncooked meat of infected bats and chimpanzees account for oral transmission in Africa. Another source is through contact with contaminated fomites and poor hygiene in hospitals (Kourtis et al, 2015; Singh et al, 2017; Hasan et al, 2019; Ebola Primer, 2020).

6.3 SYMPTOMS

The disease displays symptoms that are often on the lines of malaria, dengue fever, and typhoid. Fever, muscle pain, vomiting, diarrhea and headache are seen in the starting phase. There is also internal and external bleeding with hemorrhagic rash are later stages. EVD is typically manifested as bleeding from body apertures.

The incubation period ranges from 2 to 21 days with symptoms appearing 8–11 days post-infection. 85–95% of the patients show $>38^{\circ}$ C; 85–95% of the patients report tiredness; 52–74% show headaches; 56–58% of the patients complain of sore throat and difficulty in swallowing (dysphagia). In the advanced cases, 62–68% of patients of abdominal pain; 84–86% show diarrhea; 50–79% show vomiting and nausea.

In the late stages, there are hemorrhagic manifestations such as conjunctival bleeding, hematuria, contusion, melena, petechiae, and intraperitoneal bleeding. This progresses to confusion, dehydration, hypotension, multi-organ failure, and death. Another manifestation is maculopapular exanthema: a rash seen in 25–52% of patients during the 5th to 7th day of the disease (Kourtis et al, 2015; Singh et al, 2017; Hasan et al, 2019; Ebola Primer, 2020).

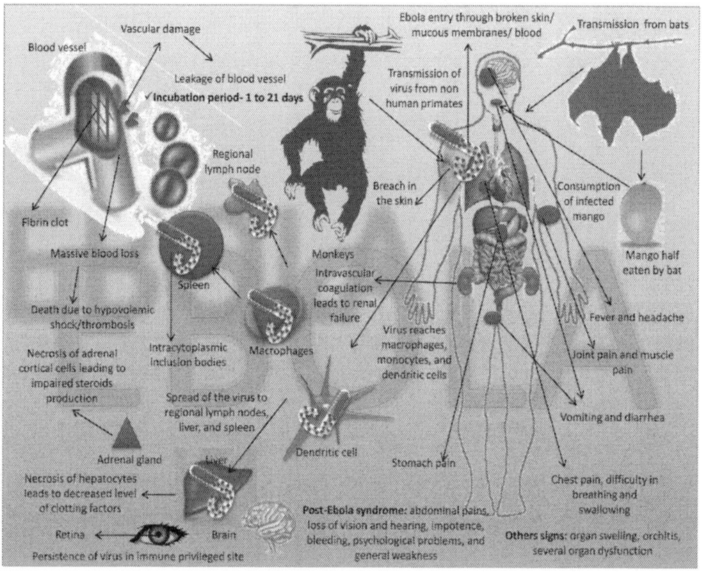

Figure 6.3. The symptoms of Ebola virus infection (Singh et al, 2017).

Additionally, survivors also report symptoms to include musculoskeletal pain in 70%; headaches in 48%, and ocular problems in 14%. 59% of the survivors showed cough, fever, appetite loss, vomiting, weight loss, itching, sneezing, earache, tremors, breathlessness, rash, and hiccups.

6.4 TREATMENT

A regime in terms of antivirals is not yet available for Ebola Virus Disease. The treatment involves the use of intravenous fluids or oral rehydration to compensate for the loss in blood volume due to bleeding. Analgesics can ease the body pain and blood transfusion would be required if the bleeding is in excess. The use of anti-emetics and anti-diarrheal medicines can address excessive vomiting and diarrhea. Appropriate antibiotics are required if secondary bacterial infections are suspected.

Studies are ongoing to develop vaccines, drugs, and antibody-based therapies:

Examples include:

- FAVIPIRAVIR (T-705) (Fujifilm Holding Corp) is a nucleotide analog that prevents viral replication: Phase II (NCT02329054): JIKI; NCT02662855.

- Genetic analysis has shown that the virulence of the Ebola virus depends on mitogen-activated protein kinase (MAPK), phosphoinositide 3-kinases (PI3Ks), and calcium/calmodulin kinases. Studies on cell lines have shown promise using inhibitors to the aforementioned molecules. For example: SB202190 (p38 MAPK inhibitor) lowered the cytokine storm induced by the virus in human monocyte-derived dendritic cells. LY294002 (PI3Ks inhibitor) and KN-93(CAMK2 inhibitor) lowered the viral infection in Vero E6 cells.
- Gam Evac-Combi vaccine induced immunity in all volunteers in phases I and II trails.
- The Sierra Leone Trial to Introduce a Vaccine against Ebola (STRIVE) program by the College of Medicine and Allied Health Sciences, University of Sierra Leone, the Sierra Leone, Ministry of Health and Sanitation, and the CDC has developed a promising VSVZEBOV vaccine. It is undergoing phase 2/ 3 trials.

6.5 PREVENTION

This is the most important aspect that necessitates that the people, medical centers, and the governments come together to prevent the incidence and spread of the disease. A major point here would be to prevent contact with an infected individual or their body fluids. This becomes important in Africa given that funeral ceremonies involve direct contact. Another important point is to prevent the handling and consumption of bush meat. A major section of the society under threat is the health care personnel who are exposed to infected patients and their fluids/clothes. Hence, all health care providers need to use personal protective equipment such as gowns, face masks, goggles, and gloves. The WHO advises abstinence or safe sex practices for survivors of the disease for at least 12 months from they showed symptoms and till they have been tested twice (Singh et al, 2017; Hasan et al, 2019; Ebola Primer, 2020).

6.6 FINAL POINTS

The Ebola epidemic ravaging parts of West Africa is the most severe acute public health emergency seen in modern times. Never before in recorded history has a biosafety level four pathogen infected so many people so quickly, over such a broad geographical area, for so long" (WHO 2014).

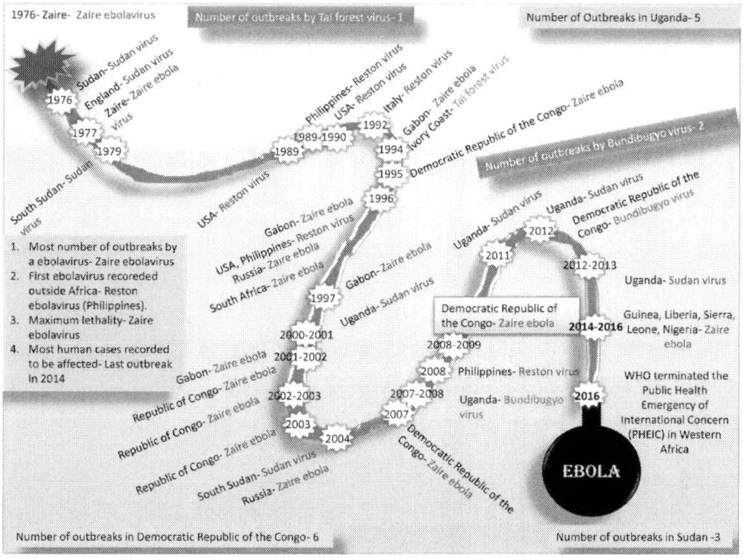

Figure 6.4. The spread of the Ebola virus infection over time (Singh et al, 2017).

These words call for utmost caution and most importantly education of the public regarding the transmission and prevention of the Ebola virus. Careful monitoring and planning can help check outbreaks that have ravaged lives and economies. Apart from proper surveillance, an important angle regarding mental health also arises. This makes the need for counseling be available to the survivors, their families, and also the healthcare professionals.

6.7 REFERENCES

1. Ebola virus disease. Nat Rev Dis Primers 6, 14 (2020). https://doi.org/10.1038/s41572-020-0154-4.
2. Kourtis AP, Appelgren K, Chevalier MS, McElroy A (2015) Ebola Virus Disease: Focus on Children. Pediatr Infect Dis J. 34(8):893-897. doi:10.1097/INF.0000000000000707.
3. Hasan S, Ahmad SA, Masood R, Saeed S (2019) Ebola virus: A global public health menace: A narrative review. J Family Med Prim Care. 8(7):2189-2201. doi:10.4103/jfmpc.jfmpc_297_19.
4. World Health Organization. Ebola virus disease – Democratic Republic of the Congo. https://www.who.int/csr/don/11-June-2020-ebola-drc/en/
5. Singh RK, Dhama K, Malik YS, Ramakrishnan MA, Karthik K, Khandia R, et al. (2017) Ebola Virus – Epidemiology, Diagnosis, and Control: Threat to Humans, Lessons Learnt, and Preparedness Plans – An Update On Its 40 Year's Journey. Veterinary Quarterly. 37(1):98-135. Doi 10.1080/01652176.2017.1309474.
6. Chan M. (2014) Ebola Virus Disease in West Africa--No Early End to The Outbreak. N Engl J Med. 371(13):1183-1185.
7. WHO Ebola Response Team (2014) Ebola Virus Disease in West Africa The First 9 Months of the Epidemic and Forward Projections. N Engl J Med. 371:1481–1495.

7

It Takes a Community to Raise the Prevalence of a Zoonotic Pathogen

Dustin Brisson,[1] **Catherine Brinkley,**[1] **Parris T. Humphrey,**[1,2] **Brian D. Kemps,**[1] **and Richard S. Ostfeld**[3]

[1]Department of Biology, University of Pennsylvania, Leidy Laboratories 209, 433 South University Avenue, Philadelphia, PA 19104-6018, USA

[2]Department of Ecology and Evolution, University of Arizona, BioSciences West room 310, 1041 E. Lowell St., Tucson, AZ, USA

[3]Cary Institute of Ecosystem Studies, Box AB, Millbrook, NY 12545, USA

ABSTRACT

By definition, zoonotic pathogens are not strict host-species specialists in that they infect humans and at least one nonhuman reservoir species. The majority of zoonotic pathogens infect and are amplified by multiple vertebrate species in nature, each of which has a quantitatively different impact on the distribution and abundance of the pathogen and thus on disease risk. Unfortunately, when new zoonotic pathogens emerge, the dominant

Citation: It Takes a Community to Raise the Prevalence of a Zoonotic Pathogen. Dustin Brisson, Catherine Brinkley, Parris T. Humphrey, Brian D. Kemps, and Richard S. Ostfeld. Hindawi Publishing Corporation Interdisciplinary Perspectives on Infectious Diseases Volume 2011, Article ID 741406, 6 pages doi:10.1155/2011/741406

response by public health scientists is to search for a few, or even the single, most important reservoirs and to ignore other species that might strongly influence transmission. This focus on the single "primary" reservoir host species can delay biological understanding, and potentially public health interventions as species important in either amplifying or regulating the pathogen are overlooked. Investigating the evolutionary and ecological strategy of newly discovered or emerging pathogens within the community of potential and actual host species will be fruitful to both biological understanding and public health.

INTRODUCTION

Zoonotic pathogens—those transmitted among nonhuman vertebrates that can infect humans—represent an increasingly important threat to human health [1, 2]. Diseases caused by zoonotic pathogens are twice as likely as strictly human pathogens to be classified as emerging or reemerging [1].

When a zoonotic disease emerges, it is imperative that public health specialists determine the underlying cause, including identifying the pathogen and its reservoirs in nature. Identifying reservoirs, the host species that maintain the pathogen in nature, can be key to interventions to reduce disease risk but rarely are effective reservoir-based interventions realized [3]. In this essay we argue that these efforts can be substantially improved by explicit recognition that zoonotic pathogens are regulated by multiple reservoir and nonreservoir hosts that must be considered in a community context.

In some cases, emerging zoonoses can be traced to a single-reservoir host species. For example, a recent outbreak of the Sin Nombre virus, a deadly member of the New World Hantavirus family, caused severe morbidity, mortality, and widespread panic in the Southwestern US in the early 1990s.

Nearly all cases can be traced back to contact with contaminated urine or feces from the deer mouse (*Peromyscus maniculatus*), the primary reservoir species of the Sin Nombre virus. Sin Nombre virus and other related Hantavirus species tend to specialize and coevolve with a specific species of mouse [4–6].

Strict specialization of emerging zoonotic pathogens on single host species, however, appears to occur in only a minority of cases. Far more frequently, emerging zoonotic pathogens use multiple reservoir hosts. One recent example is the West Nile virus (WNV) epidemic in the United States, which has caused widespread morbidity, mortality, and concern. Unlike Hantavirus species, WNV naturally infects a tremendous array of vertebrate species including at least 300 bird species and some mammals [7–9].

In part owing to the pervasive assumption that zoonotic pathogens specialize on particular reservoir hosts, much of the public and academic discourse on WNV focuses on corvids (especially American crows) because they die conspicuously after infection and are considered harbingers of an approaching disease outbreak [8, 10–13]. This focus on corvids leads to the misperception that crows are the primary or principal reservoir of WNV [9, 14–18] when in fact they are relatively unimportant in WNV maintenance and amplification (increasing abundance in nature) due to their relative rarity at some sites and the preference of mosquitoes for other bird species such as American robins [8].

The notion that each zoonotic pathogen has a primary or principal vertebrate reservoir species permeates the scientific literature. Over two-thirds of the publications in PubMed that investigate zoonotic pathogens and their natural reservoirs focus on only one vertebrate species (Figure 1(a)).

By contrast, over two-thirds of pathogens classified as zoonoses infect multiple nonhuman vertebrate species [1, 19–21] (Figure 1(b)). Some of the studies that focus on a single host species are intended to provide specific information that can be used for later, more holistic assessments of zoonotic risk.

However, in reality, the more holistic syntheses are rarely performed, preserving the paradigm that zoonotic pathogens have a primary reservoir host species. Broader syntheses are important as every infected vertebrate species in an ecological community can affect the overall reproductive success and abundance of zoonotic microbes.

Further, every member of the focal community, infected or not, has the potential to affect human disease risk by contributing to the transmission dynamics and overall abundance of the pathogen in nature. Here, we advocate a more inclusive approach to understanding the underlying ecological basis of variation in zoonotic disease risk whenever time and financial constraints allow.

Both the basic understanding of the causes of variable risk and the creation of effective disease management plans would benefit from explicit recognition that zoonotic pathogens are typically regulated by a community of hosts.

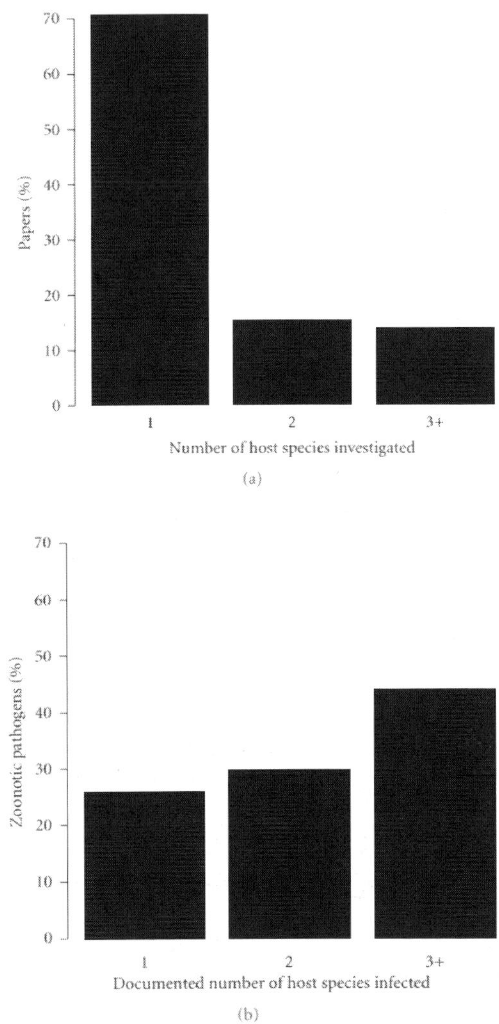

Figure 1. Although the vast majority of zoonotic disease scientific publications focus on a single vertebrate host species, the majority of zoonotic pathogens infect multiple host species. (a) Over 70% of the 1672 publications listed on PubMed that focus on a zoonotic pathogen investigated a single vertebrate host species. (b) By contrast, nearly 74% of the 865 zoonotic pathogens with documented host-species ranges infect multiple host species [1, 19–21].

Public health measures to mitigate zoonotic disease incidence often focus on preventing transmission between nonhuman vertebrates and humans or controlling infection in the nonhuman vertebrate hosts. The focus on a single reservoir species obscures the basic biological features of zoonotic pathogens. The 2003 SARS outbreak in China serves as an example. At the outset of this epidemic, epidemiologists set as a top priority the need "to identify the animal reservoir" [22]. The assumption that the SARS coronavirus (CoV) had one primary reservoir host species was so pervasive that the Chinese government mandated the culling of ~10,000 palm civets (*Paguma larvata,* see Supplemental Figure 1 in Supplementary Material available online at doi:10.1155/2011/741406) after only six individual civets from a live-animal market tested positive [23]. Subsequent investigations suggested that the infected civets likely acquired the virus in a live-animal market from another species [24]. The search for natural reservoirs of SARS-CoV has revealed just how nonspecialized this pathogen can be. Several species of horseshoe bats (*Rhinolophus* spp., Supplemental Figure 1) have been implicated as reservoirs of SARS-like CoV [23–25], but their roles in amplifying the pathogen and in transmitting it interspecifically remain obscure. Although several bat species clearly harbor active infections, with six species having >15% prevalence, none of the bats examined so far appear to transmit the virus through the respiratory tract, thought to be the primary or sole mode of transmission [25]. Active SARS-like CoV infections have been discovered in at least six nonbat species at a prevalence greater than 15%, but their roles in maintaining and transmitting SARS-like CoV similarly are unknown. Further studies are needed to quantify the roles that each host species plays in amplifying and transmitting the pathogen among species and to humans.

The distribution and abundance of multihost zoonotic pathogens—and thus disease risk—is dependent on the distribution, abundance, and behaviors of *many* animal species with which the pathogen interacts, not only the most efficient or primary reservoir species. Animals within each of these species act as the environments within which zoonotic pathogens can propagate and from which they can disperse. The total abundance of a zoonotic pathogen population is a direct function of its replication, transmission, and colonization success in each environment it colonizes, or attempts to colonize. Source-sink theory in ecology posits that organisms exist in some habitats (sinks) only because of immigration from other habitats (sources) [26]. In sink habitats, the rate of reproduction does not meet or exceed that of mortality such that sinks are not self-sustaining but instead require immigration. Sources are self-sustaining environments that can provide dispersers that might colonize sinks. Zoonotic pathogen abundance and distribution is regulated by both source and sink environments, dispersal rates between these environments, and the sizes and positions of source and sink environments. Pathogens exist in analogous landscapes consisting of multiple reservoir hosts of various quality (sources) and nonreservoirs (sinks), with the enrichment that each vertebrate species is mobile so that the landscape is constantly shifting. Hence, discovering that a species is infected with a pathogen does not necessarily mean that species is a pathogen reservoir. It could be a "sink" or spillover host the net effect of which is to *reduce* pathogen replication. Just as several habitat types can differ in their contributions to overall population growth of an animal occupant, so too can different reservoirs differ in how they contribute to the force of infection.

Below we explore the ecology of zoonotic pathogens as the complex of interactions with the animal community and relate the resulting evolutionary fitness of the pathogens to human disease risk. The evolutionary fitness, and thus the abundance, of a pathogen is dependent on a variety of factors that emerge from both the direct and indirect interactions between the pathogen and features of their environment. In this paper, we explore the effects of the direct and indirect interactions between zoonotic pathogens and (1) animal species that amplify pathogen abundance in nature; (2) animal species that negatively impact the abundance of pathogens; (3) animal species that facilitate pathogen amplification but that are not directly involved in the transmission cycle. Additionally, we explore the effects of interaction among vertebrate host species in the community on pathogen fitness. Within this framework, the evolutionary and ecological strategies of a pathogen that promote its continual existence directly affect human disease risk. Although this commentary is focused explicitly on vertebrate host species, the general argument may also be applicable to invertebrate vectors, an idea that should be explored in future work.

ECOLOGY OF EVOLUTIONARILY SUCCESSFUL ZOONOTIC PATHOGENS

The fitness of zoonotic pathogens is controlled by the quantity and quality of their environments, namely, each of the animal populations with which it interacts. Reservoir species are ecological sources of the pathogen in nature. These species act as "high-quality patches" that make a positive contribution to the overall abundance of the zoonotic pathogen [26, 27]. A high rate of secondary cases caused by a typical infected individual—thus increasing the abundance of the pathogen—is a hallmark of a reservoir species [28–30]. All evidence suggests that the majority of zoonotic pathogens use multiple reservoir species. We suggest that this community of reservoir species collectively maintains or increases the abundance of the zoonotic pathogen.

While species that contribute positively to the abundance of the zoonotic pathogen and to human disease risk are all qualitatively reservoirs of the pathogen, each reservoir species has a quantitatively different impact on pathogen abundance and human disease risk. The quantitative effect of a reservoir species on the abundance of a zoonotic pathogen is a function of five commonly used parameters [28, 31] that vary depending on the reservoir species and on the zoonotic pathogen: (1) host-species density, (2) host infection prevalence, (3) transmission rate, (4) transmission duration, and (5) the intra- and interspecific connectivity among hosts in the community. Pathogen population growth is the result of each of these species-specific parameter values integrated across all animal species in the community. Estimates of these parameters in natural settings will allow public health officials to determine the most effective set of species to target to substantially decrease disease risk, regardless of whether they are the primary reservoir species.

Host Density and Infection Prevalence

Host density and infection prevalence are the most commonly assessed parameters of reservoir species in nature. These parameters are rarely measured in species that are difficult to work with or are inconspicuous. For example, the power of rodents as reservoir species of the cowpox virus (*Orthopoxvirus bovis*) has been hidden behind the highly conspicuous domestic cow—the presumed "primary" reservoir species [32, 33]. Cows had been considered the primary reservoir of cowpox for at least five hundred years until urban centers (devoid of cattle) experienced a dramatic increase in human cases in the middle of the 20th century. Cows show visual signs of infection and have traditionally been associated with infected humans. In the last few decades, small rodents were identified as a natural reservoir host species [34]. New evidence suggests that the community of reservoir species is even more complex as new reservoir species continue to surface [35, 36].

Transmission Rate and Duration

Transmission rate and transmission duration determine the maximum number of susceptible individuals that can be infected as a result of a single infected individual [37–40]. The transmission strategies of zoonotic pathogens often depend on the species host that is infected. For example,

Nipah virus infections are highly virulent in pigs and are characterized by a high level of excretion and a short transmission duration [41], while the virus in grey-headed fruit bats (*Pteropus poliocephalus*) is less virulent with low-level excretion over long time periods [42]. Virulence of a pathogen within each of its hosts has been shaped by trade-offs between the need to increase in numbers within the infected host and the need to disperse to new uninfected hosts [43]. Physiological and immunological differences between hosts will inevitably result in host-specific optima. Consequently, reservoir species often differ quantitatively in their effect on the magnitude and duration of pathogen transmission, and thus human disease risk.

Inter- and Intraspecific Connectivity

The evolutionary success for a zoonotic pathogen is a function of the average number of susceptible animals that become infected as a result of a single infected animal [38, 39]. The number of susceptible animals that become infected is dependent in part on the connectivity among hosts with regard to pathogen transmission. High connectivity among individuals within a reservoir species or among reservoir species contributes to increased pathogen abundance in the host community. In fact, interspecies connectivity (among reservoir species) can be essential to pathogen success. For example, cowpox cannot be maintained in a rodent community consisting solely of wood mice (*Apodemus sylvaticus*) even though wood mice are easily infected and transmit cowpox. Intraspecies connectivity among wood mice is limited such that extinction of the cowpox virus occurs shortly after introduction into a wood mouse population. However, connectivity among reservoirs is sufficient to maintain cowpox when the vertebrate community includes wood mice, bank voles (*Myodes glareolus*), and field voles (*Microtus agrestis*) [44].

DILUTING SPECIES

Zoonotic pathogens, like all species, experience environments to which they are poorly suited and in which they rarely persist. These low-quality habitats, or "dilution species," negatively impact the abundance of pathogens and thus can decrease human disease risk [45]. Disease dilution can occur by several mechanisms, at least three of which are common in nature [46]. The first is by reducing rates of encounter between pathogens and reservoir species. For example, cattle used as zoo prophylaxis to deflect malaria-inducing mosquito bites from humans reduce vector-host encounters that would allow pathogen amplification are reduced [47–49]. Similarly, the presence of predators can reduce the activity rates and home range sizes of small-mammal reservoir species, thereby reducing their encounter rates with other susceptible hosts or vectors [50, 51]. Second, the presence of species other than reservoir species can dilute disease risk by regulating abundance of the reservoir species. For example, competitors can strongly regulate the abundance of a reservoir species, reducing the overall abundance of a reservoir species and thus the pathogen [52]. Third, in vector-borne zoonoses, the presence of a variety of species can act to regulate the abundance of the vector itself, thereby modulating pathogen transmission. For example, some hosts for the tick vector of the Lyme disease bacterium, *Borrelia burgdorferi,* are poor quality hosts for ticks, killing the majority that attempt to take a bloodmeal. The result is a strong reduction in vector abundance compared with host communities in which species of low quality for ticks are rare or absent [53].

Antibody surveys are commonly used to assess disease prevalence in wildlife populations. However, serologic evidence of infection may be common in both reservoir and dilution species as both reservoir and dilution species are often exposed and may produce antibodies to the pathogen. Disease control measures aimed at species with high seroprevalence may inadvertently target a dilution species and *increase* disease risk. Many mammalian species, such as raccoons, have high seroprevalence for WNV although there is no demonstration to date that they transmit the virus to feeding vectors [54]. Mammals may reduce the prevalence of WNV by attracting mosquito bites that would have otherwise infected a reservoir species such as some passerine birds [55].

VECTOR AMPLIFIERS

Vector-borne diseases comprise some of the most virulent human diseases. Plague, yellow fever, and tularemia cause severe, debilitating symptoms and are still regularly transmitted to naïve hosts due to the activity of vectors. In fact, transmission of mosquito-borne pathogens may even be enhanced by debilitating symptoms as immobilized hosts are poorer at avoiding mosquito bites than are asymptomatic hosts, resulting in longer feeding bouts that increase the probability of acquiring blood-borne pathogens [43]. Although a healthy vector population is imperative for vector-borne disease connectivity, the species that are important for maintaining the vector population often differ from those that are important for maintaining the pathogen. For example, in the far Western United States, the western fence lizard (*Sceloporus occidentalis*) is important in maintaining vector tick populations, but rodents are necessary to infect ticks with Lyme disease spirochetes [56, 57]. In such systems, controlling a nonreservoir host important in maintaining vectors might be more effective than controlling reservoir hosts that are important in transmitting the pathogen.

COMMUNITY EFFECTS

Classical ecological interactions among organisms such as competition, predation, and behavior are important regulators of zoonotic pathogen abundance and disease risk. Species that play no direct role in increasing or diluting pathogen abundance can affect pathogen abundance by regulating the density of reservoir species. For example, the density of many rodent populations is regulated by resource abundance. Recent studies show that the density, distribution, and timing of acorn production from oak trees affect Lyme disease risk by controlling the densities of white-footed mice and chipmunks—reservoir host species for Lyme disease bacteria in the Northeastern US—although oak trees themselves have no direct interaction with *B. burgdorferi* [58].

SUMMARY AND CONCLUSIONS

The majority of zoonotic pathogens infect, and are amplified by, multiple vertebrate species despite the tendency of much of the research community to focus efforts on a single "primary" reservoir species. Nonhuman vertebrate species with which a zoonotic pathogen interacts behave as environments of differing quality, such that each environment has a quantitatively different effect on pathogen prevalence and abundance and thus on disease risk. We advocate a new approach in response to newly emerging zoonotic diseases that focuses on the community of species that directly or indirectly affects the abundance of a zoonotic pathogen. This set of species is likely to include multiple reservoirs and might include dilution species as well as those that regulate reservoir or dilution species. Although targeting a set of species is more difficult than pursuing only one "primary" reservoir species, the community-based approach will help deter false leads and ineffective public health interventions that are based on an oversimplified ecology of the zoonotic pathogen. Pursuing a more inclusive set of vertebrates involved in zoonotic transmission will also be more likely to uncover cryptic yet critical interactions affecting risk.

REFERENCES

1. Taylor LH, Latham SM, Woolhouse MEJ. Risk factors for human disease emergence. *Philosophical Transactions of the Royal Society B*. 2001;356(1411):983–989.
2. Jones KE, Patel NG, Levy MA, et al. Global trends in emerging infectious diseases. *Nature*. 2008;451(7181):990–993.
3. Peters CJ. Resolved: emerging infections of humans can be controlled by ecological interventions. In: Ostfeld RS, Keesing F, Eviner VT, editors. *Infectious Disease Ecology: Effects of Ecosystems on Disease and of Disease on Ecosystems*. Princeton, NJ, USA: Princeton University Press; 2008. pp. 423–440.
4. Monroe MC, Morzunov SP, Johnson AM, et al. Genetic diversity and distribution of Peromyscus-borne hantaviruses in North America. *Emerging Infectious Diseases*. 1999;5(1):75–86.
5. Hawes S, Seabolt JP. Hantavirus. *Clinical Laboratory Science*. 2003;16(1):39–42.
6. Khaiboullina SF, Morzunov SP, Jeor SCS. Hantaviruses: molecular biology, evolution and pathogenesis. *Current Molecular Medicine*. 2005;5(8):773–790.
7. Hayes EB, Komar N, Nasci RS, Montgomery SP, O'Leary DR, Campbell GL. Epidemiology and transmission dynamics of West Nile virus disease. *Emerging Infectious Diseases*. 2005;11(8):1167–1173.
8. Kilpatrick AM, Daszak P, Jones MJ, Marra PP, Kramer LD. Host heterogeneity dominates West Nile virus transmission. *Proceedings of the Royal Society B*. 2006;273(1599):2327–2333.
9. Pollock CG. West Nile virus in the Americas. *Journal of Avian Medicine and Surgery*. 2008;22(2):151–157.
10. Garvin MC, Tarvin KA, Smith J, Ohajuruka OA, Grimes SD. Patterns of West Nile virus infection in Ohio blue jays: implications for initiation of the annual cycle. *American Journal of Tropical Medicine and Hygiene*. 2004;70(5):566–570.
11. Reisen WK, Barker CM, Carney R, et al. Role of corvids in epidemiology of West Nile virus in Southern California. *Journal of Medical Entomology*. 2006;43(2):356–367.
12. Ward MP, Raim A, Yaremych-Hamer S, Lampman R, Novak RJ. Does the roosting behavior of birds affect transmission dynamics of West Nile virus? *American Journal of Tropical Medicine and*

Hygiene. 2006;75(2):350–355.

13. Koenig WD, Hochachka WM, Zuckerberg B, Dickinson JL. Ecological determinants of American crow mortality due to West Nile virus during its North American sweep. *Oecologia*. 2010;163(4):903–909.
14. Brault AC, Langevin SA, Bowen RA, et al. Differential virulence of West Nile strains for American crows. *Emerging Infectious Diseases*. 2004;10(12):2161–2168.
15. Weingartl HM, Neufeld JL, Copps J, Marszal P. Experimental West Nile virus infection in blue jays (Cyanocitta cristata) and crows (Corvus brachyrhynchos) *Veterinary Pathology*. 2004;41(4):362–370.
16. Langevin SA, Brault AC, Panella NA, Bowen RA, Komar N. Variation in virulence of West Nile virus strains for house sparrows (Passer Domesticus) *American Journal of Tropical Medicine and Hygiene*. 2005;72(1):99–102.
17. Bowen RA, Nemeth NM. Experimental infections with West Nile virus. *Current Opinion in Infectious Diseases*. 2007;20(3):293–297.
18. David ST, Mak S, MacDougall L, Fyfe M. A bird's eye view: using geographic analysis to evaluate the representativeness of corvid indicators for West Nile virus surveillance. *International Journal of Health Geographics*. 2007;6:p. 3.
19. Cleaveland S, Laurenson MK, Taylor LH. Diseases of humans and their domestic mammals: pathogen characteristics, host range and the risk of emergence. *Philosophical Transactions of the Royal Society B*. 2001;356(1411):991–999.
20. Woolhouse MEJ, Gowtage-Sequeria S. Host range and emerging and reemerging pathogens. *Emerging Infectious Diseases*. 2005;11(12):1842–1847.
21. Friend M. Disease emergence and resurgence: the wildlife-human connection. U.S. Dept. of the Interior, U.S. Geological Survey, Reston, Va, USA, 2006.
22. Altman LK. *Experts Urge Tightening Of Safeguards In SARS Labs*. New York, NY, USA: New York Times; 2003.
23. Wang LF, Eaton BT. Bats, civets and the emergence of SARS. *Current Topics in Microbiology and Immunology*. 2007;315:325–344.
24. Shi Z, Hu Z. A review of studies on animal reservoirs of the SARS coronavirus. *Virus Research*. 2008;133(1):74–87.

25. Li W, Shi Z, Yu M, et al. Bats are natural reservoirs of SARS-like coronaviruses. *Science*. 2005;310(5748):676–679.
26. Pulliam HR. Sources, sinks and population regulation. *American Naturalist*. 1988;132(5):652–661.
27. Holt RD. Population dynamics in two-patch environments: some anomalous consequences of an optimal habitat distribution. *Theoretical Population Biology*. 1985;28(2):181–208.
28. Anderson RM, May RM. Population biology of infectious diseases: part I. *Nature*. 1979;280(5721):361–367.
29. Nishiura H, Inaba H. Discussion: emergence of the concept of the basic reproduction number from mathematical demography. *Journal of Theoretical Biology*. 2007;244(2):357–364.
30. Hartemink NA, Randolph SE, Davis SA, Heesterbeek JAP. The basic reproduction number for complex disease systems: defining R_0 for tick-borne infections. *American Naturalist*. 2008;171(6):743–754.
31. May RM, Anderson RM. Population biology of infectious diseases: part II. *Nature*. 1979;280(5722):455–461.
32. Jenner E. *An Inquiry into the Causes and Effects of the Variolae Vaccinae, a Disease Discovered in Some of the Western Counties of England, Particularly Gloucestershire, and Known by the Name of the Cow Pox*. London, UK: Sampson Low; 1798.
33. Baxby D. Is cowpox misnamed? A review of 10 human cases. *British Medical Journal*. 1977;1(6073):1379–1381.
34. Miller CG, Shchelkunov SN, Kotwal GJ. The cowpox virus-encoded homolog of the vaccinia virus complement control protein is an inflammation modulatory protein. *Virology*. 1997;229(1):126–133.
35. Charbonnel N, Deter J, Chaval Y, et al. Serological evidence of viruses naturally associated with the montane water vole (*Arvicola scherman*) in eastern France. *Vector-Borne and Zoonotic Diseases*. 2008;8(6):763–767.
36. Kurth A, Wibbelt G, Gerber HP, Petschaelis A, Pauli G, Nitsche A. Rat-to-elephant-to-human transmission of cowpox virus. *Emerging Infectious Diseases*. 2008;14(4):670–671.
37. Dublin L, Lotka A. On the true rate of natural increase. *Journal of the American Statistical Association*. 1925;150:305–339.
38. Diekmann O, Heesterbeek JA, Metz JA. On the definition and the computation of the basic reproduction ratio R_0 in models for infectious

diseases in heterogeneous populations. *Journal of Mathematical Biology*. 1990;28(4):365–382.

39. Anderson RM, May RM. *Infectious Diseases of Humans: Transmission and Control*. Oxford, UK: Oxford University Press; 1991.
40. Heesterbeek JAP. A brief history of R0 and a recipe for its calculation. *Acta Biotheoretica*. 2002;50(3):189–204.
41. Middleton DJ, Westbury HA, Morrissy CJ, et al. Experimental Nipah virus infection in pigs and cats. *Journal of Comparative Pathology*. 2002;126(2-3):124–136.
42. Middleton DJ, Morrissy CJ, van der Heide BM, et al. Experimental Nipah Virus infection in pteropid bats (*Pteropus poliocephalus*) *Journal of Comparative Pathology*. 2007;136(4):266–272.
43. Ewald PW. *Evolution of Infectious Disease*. New York, NY, USA: Oxford University Press; 1996.
44. Chantrey J, Meyer H, Baxby D, et al. Cowpox: reservoir hosts and geographic range. *Epidemiology and Infection*. 1999;122(3):455–460.
45. Ostfeld RS, Keesing F. Biodiversity and disease risk: the case of Lyme disease. *Conservation Biology*. 2000;14(3):722–728.
46. Keesing F, Holt RD, Ostfeld RS. Effects of species diversity on disease risk. *Ecology Letters*. 2006;9(4):485–498.
47. Sota T, Mogi M. Effectiveness of zooprophylaxis in malaria control: a theoretical inquiry, with a model for mosquito populations with two bloodmeal hosts. *Medical and Veterinary Entomology*. 1989;3(4):337–345.
48. Dobson A, Cattadori I, Holt RD, et al. Sacred cows and sympathetic squirrels: the importance of biological diversity to human health. *PLoS Medicine*. 2006;3(6, article e231)
49. Mahande A, Mosha F, Mahande J, Kweka E. Feeding and resting behaviour of malaria vector, *Anopheles arabiensis* with reference to zooprophylaxis. *Malaria Journal*. 2007;6:p. 100.
50. LoGiudice K, Duerr STK, Newhouse MJ, Schmidt KA, Killilea ME, Ostfeld RS. Impact of host community composition on lyme disease risk. *Ecology*. 2008;89(10):2841–2849.
51. Dizney LJ, Ruedas LA. Increased host species diversity and decreased prevalence of Sin Nombre virus. *Emerging Infectious Diseases*. 2009;15(7):1012–1018.

52. Suzán G, Marcé E, Giermakowski JT, et al. Experimental evidence for reduced rodent diversity causing increased hantavirus prevalence. *PLoS ONE*. 2009;4(5, article e5461)
53. Keesing F, Brunner J, Duerr S, et al. Hosts as ecological traps for the vector of Lyme disease. *Proceedings of the Royal Society B*. 2009;276(1675):3911–3919.
54. Root JJ, Hall JS, McLean RG, et al. Serologic evidence of exposure of wild mammals to flaviviruses in the central and eastern United States. *American Journal of Tropical Medicine and Hygiene*. 2005;72(5):622–630.
55. Ezenwa VO, Godsey MS, King RJ, Guptill SC. Avian diversity and West Nile virus: testing associations between biodiversity and infectious disease risk. *Proceedings of the Royal Society B*. 2006;273(1582):109–117.
56. Lane RS, Piesman J, Burgdorfer W. Lyme borreliosis: relation of its causative agent to its vectors and hosts in North America and Europe. *Annual Review of Entomology*. 1991;36(1):587–609.
57. Swei A, Ostfeld RS, Lane RS, Briggs CJ. Impact of the experimental removal of lizards on Lyme disease risk. *Proceedings of the Royal Society B*. 2011;278(1720):2970–2978.
58. Ostfeld RS, Canham CD, Oggenfuss K, Winchcombe RJ, Keesing F. Climate, deer, rodents, and acorns as determinants of variation in lyme-disease risk. *PLoS Biology*. 2006;4(6):p. e145.

8

Multiple Control Strategies for Prevention of Avian Influenza Pandemic

Roman Ullah,[1,2] Gul Zaman,[3] and Saeed Islam[2]

[1]Department of Mathematics, Bacha Khan University, Charsadda 24420, Pakistan

[2]Department of Mathematics, Abdul Wali Khan University Mardan, Mardan 23200, Pakistan

[3]Department of Mathematics, University of Malakand, Chakdara 23101, Pakistan

ABSTRACT

We present the prevention of avian influenza pandemic by adjusting multiple control functions in the human-to-human transmittable avian influenza model. First we show the existence of the optimal control problem; then by using both analytical and numerical techniques, we investigate the cost-effective control effects for the prevention of transmission of disease. To do this, we use three control functions, the effort to reduce the number of

Citation: Multiple Control Strategies for Prevention of Avian Influenza Pandemic Roman Ullah, Gul Zaman, and Saeed Islam. Hindawi Publishing Corporation e Scientific World Journal Volume 2014, Article ID 949718, 9 pages http://dx.doi.org/10.1155/2014/949718

contacts with human infected with mutant avian influenza, the antiviral treatment of infected individuals, and the effort to reduce the number of infected birds. We completely characterized the optimal control and compute numerical solution of the optimality system by using an iterative method.

INTRODUCTION

Highly pathogenic avian influenza is a zoonotic disease caused by H5N1 virus and it has a devastating impact on poultry causing above 90 percent mortality within 48 hours of infection [1]. The disease also spreads to humans through the direct contact with the infected poultry. Despite all advances of medical, avian influenza still poses a significant threat to health and economy of the society. The laboratory confirmed that the fatality rate in human case is about 60 percent [2].

Several reports have so far been made of possible coinfection of humans with an H5N1 strain and a human strain [3, 4]. One of the coinfection reports is of a Turkish family cluster. In late December 2005 and early January 2006, a cluster of 8 confirmed highly pathogenic avian influenza (H5N1) cases was detected in Dogubayazit district in eastern Turkey [5]. All confirmed case patients were hospitalized after onset of symptoms. Four of them were died and the other four recovered. Another cluster of 8 cases of HPAI (H5N1) was detected in the village of Kubu Sambilang, Samatra, Indonesia, in April-May 2006 [6]. Only one of them recovered and the remaining seven members died. Though the human-to-human transmission of the disease is rare, its potential to change into an extremely high virulent human-to-human transmittable pandemic strain is the real danger to human health. Due to this threat, some international organizations have developed and began to implement different strategies (biosecurity, isolation, and antiviral treatment) for forestalling the onset of a pandemic. It has been estimated that a future avian influenza pandemic could cause hundreds of billions of dolors due to lost productivity, associated medical policies, and prevention policies.

Mathematical models have been widely used to evaluate the effects of control measures of avian influenza [7–9]. In 2007, Iwami et al. [10] proposed a mathematical model of the spread of avian influenza from bird population to human population. They discussed that to minimize the spread of the disease in human population, we must take the measures for the infected human with bird flu to quarantine when mutant bird flu has already occurred. Lee et al. [11] show that the optimal control strategies that rely on the use of antiviral treatment or isolation strategies can reduce the number of clinical cases; therein they emphasize isolation strategies in mitigation of pandemic of influenza particularly when the access to antiviral resources is limited. Later on Ullah et al. [12] proposed the control strategies where they focus on minimizing the impact of influenza by minimizing the vaccine wanning.

In this paper, we focus on identifying the optimal control strategies for the model developed in [13]; these controls minimize the impact of avian influenza by isolating the infected individuals, the judicious use of drug supply, and eliminating the infected birds. Recommended medicines like oseltamivir or zanamivir must be taken within two days after the appearance of symptoms. The clinically ill and infectious individuals can be isolated by reasonably effective ways to reduce the transmission of disease, like to educate them to cover their sneeze and cough, not to spit openly, and to avoid the closed contacts with others, sanitizing the rooms or equipments occupied by the patients. The diagnosed infected birds should be culled within a few hours and all the infected dead birds and other contaminated objects (faeces, blood, and feathers) must be destroyed properly (burying, incineration) as soon as possible. We can get the most effective results if we use the isolation, treatment of clinically infectious humans, and elimination of infected birds concurrently.

MODEL FRAME WORK

In this section, the compartmental model that divides the human and birds populations into two different classes presents the optimal control problem for the transmission dynamics of avian influenza. Our main aim is to show that it is possible to implement the time-dependent antiavian influenza control techniques while minimizing the cost of such measures. In order to do this, we introduce three time dependent control functions u_1, u_2, and u_3. The control $u1$ represents the effort to reduce the number of contacts with human infected by mutant avian influenza, u_2 represents the fraction of clinically infectious cases treated with antiviral per unit of time, and u_3 is the effort to reduce the number of infected birds. Note that the controls are fully effective when $ui = 1$ for $i = 1, 2, 3$, while there is no control if $u_i = 0$. We divide the total human population at time t denoted by N_h into five distinct epidemiological subclasses which are susceptible X_h, exposed E_h, clinically ill and infectious I_h, treated T_h, and recovered R_h and the birds population Nb into three distinct subclasses which are susceptible X_b, exposed E_b, and infected I_b. Taking into account the assumptions above, the dynamics of the control problem is given by

$$\frac{dX_h}{dt} = \Lambda - \mu_h X_h - (1 - u_1)\,\alpha_1 X_h I_h - \alpha_2 X_h I_b + \epsilon_h R_h,$$
$$\frac{dE_h}{dt} = (1 - u_1)\,\alpha_1 X_h I_h + \alpha_2 X_h I_b - (\mu_h + \phi_h)\,E_h,$$
$$\frac{dI_h}{dt} = \phi_h E_h - (\rho_h + \beta_h + \mu_h + u_2)\,I_h,$$
$$\frac{dT_h}{dt} = \rho_h I_h - (\gamma_h + \mu_h)\,T_h,$$
$$\frac{dR_h}{dt} = \gamma_h T_h + u_2 I_h - (\mu_h + \epsilon_h)\,R_h,$$

$$\frac{dX_b}{dt} = \pi - \mu_b X_b - (1 - u_3)\,\alpha_3 X_b I_b,$$
$$\frac{dE_b}{dt} = (1 - u_3)\,\alpha_3 X_b I_b - (\mu_b + \phi_b)\,E_b,$$
$$\frac{dI_b}{dt} = \phi_b E_b - (\beta_b + \mu_b)\,I_b, \qquad (1)$$

with the initial conditions

$$X_h(0) \geq 0, \quad E_h(0) \geq 0, \quad I_h(0) \geq 0,$$
$$T_h(0) \geq 0, \quad R_h(0) \geq 0, \quad X_b(0) \geq 0,$$
$$E_b(0) \geq 0, \quad I_b(0) \geq 0. \qquad (2)$$

The human population is recruited at a constant birth rate Λ, ϵ_h is the rate of immunity loss, ϕh is the progression rate from E_h to I_h, ρ_h represents the treatment rate of human, μ_h and μ_b are the natural death rates of humans and birds, respectively, β_h and β_b are disease induced death rates in humans and birds, respectively, π represents recruitment rate of birds population, γ_h shows recovery due to treatment, ϕ_b is progression rate of birds from E_b to I_b, and α_1, α_2, and α_3 are effective contact rates between X_h and I_h, X_h and I_b, and X_b and I_b, respectively.

The objective of our control problem is to minimize the number of clinically infectious humans and infected birds and the cost of implementing the control by using possible minimal control variables (t) for $i = 1, 2, 3$. We use the Lebesgue measurable control and define our objective functional as

$$J(u_1, u_2, u_3) = \int_0^{t_{\text{end}}} \left(A_1 I_h + A_2 I_b + \frac{1}{2}\left(C_1 u_1^2 + C_2 u_2^2 + C_3 u_3^2\right)\right) dt, \quad (3)$$

where A_1 and A_2 are positive constants to keep a balance in the size of I_h and I_b.The square of the control variables reflects the severity of the side effects of the controls. C_1, C_2, and C_3 are positive weight parameters such that 0< $\Lambda/\mu h$, 0< $\Lambda/\mu h$, and 0< $\pi/\mu b$. The objective of the optimal control problem is to seek optimal control functions ($u^*{}_1(t)$, $u^*{}_2(t)$, $u^*{}_3(t)$) such that

$$J(u_1^*, u_2^*, u_3^*) = \min_{(u_1, u_2, u_3)\in U} \{J(u_1, u_2, u_3) \mid (u_1, u_2, u_3) \in U\}, \quad (4)$$

where the control set is defined as

$$U = \{ \ = (u_1, u_2, u_3) | u_i \text{ is Lebesgue measurable on } [0, 1], \\ 0 \le u_i(t) \le 1, \ t \in [0, t_{\text{end}}], \text{ for } i = 1, 2, 3\}, \quad (5)$$

EXISTENCE OF CONTROL PROBLEM

In this section, we consider the control system (2) with initial conditions at t=0 to show the existence of the control problem. Note that, for the bounded Lebesgue measurable controls and nonnegative initial conditions, nonnegative bounded solutions to the state system exist (see [15]). To prove the existence of the solution of system (2), we can write

$$Y_t = BY + F(Y), \quad (6)$$

where

$$Y = \begin{bmatrix} X_h \\ E_h \\ I_h \\ T_h \\ R_h \\ X_b \\ E_b \\ I_b \end{bmatrix}, \quad F(Y) = \begin{bmatrix} \Lambda - (1-u_1)\alpha_1 X_h I_h - \alpha_2 X_h I_b \\ (1-u_1)\alpha_1 X_h I_h + \alpha_2 X_h I_b \\ 0 \\ 0 \\ 0 \\ \pi - (1-u_3)\alpha_3 X_b I_b \\ (1-u_3)\alpha_3 X_b I_b \\ 0 \end{bmatrix},$$

$$B = \begin{bmatrix} -\mu_h & 0 & 0 & 0 & \epsilon_h & 0 & 0 & 0 \\ 0 & -\mu_h - \phi_h & 0 & 0 & 0 & 0 & 0 & 0 \\ 0 & \phi_h & -\rho_h - \beta_h - \mu_h - u_2 & 0 & 0 & 0 & 0 & 0 \\ 0 & 0 & \rho_h & -\gamma_h - \mu_h & 0 & 0 & 0 & 0 \\ 0 & 0 & 0 & \gamma_h + u_2 & -\mu_h - \epsilon_h & 0 & 0 & 0 \\ \mu_h & 0 & 0 & 0 & 0 & 0 & 0 & 0 \\ 0 & 0 & 0 & 0 & 0 & 0 & -\mu_b - \phi_b & 0 \\ 0 & 0 & 0 & 0 & 0 & 0 & \phi_b & -\beta_b - \mu_b \end{bmatrix}, \quad (7)$$

where Y_t denotes derivative of Y with respect to time t. Equation (3) is a nonlinear system with a bounded coefficient. We set

$$D(Y) = BY + F(Y). \quad (8)$$

Now,

$$F(Y_1) - F(Y_2)$$
$$= \begin{bmatrix} (1-u_1)\alpha_1(X_{h2}I_{h2} - X_{h1}I_{h1}) + \alpha_2(X_{h2}I_{b2} - X_{h1}I_{b1}) \\ -(1-u_1)\alpha_1(X_{h2}I_{h2} - X_{h1}I_{h1}) - \alpha_2(X_{h2}I_{b2} - X_{h1}I_{b1}) \\ 0 \\ 0 \\ 0 \\ (1-u_3)\alpha_3(X_{b2}I_{b2} - X_{b1}I_{b1}) \\ -(1-u_3)\alpha_3(X_{b2}I_{b2} - X_{b1}I_{b1}) \\ 0 \end{bmatrix}, \tag{8}$$

$$\begin{aligned} &|F(Y_1) - F(Y_2)| \\ &= |(1-u_1)\alpha_1(X_{h2}I_{h2} - X_{h1}I_{h1}) + \alpha_2(X_{h2}I_{b2} - X_{h1}I_{b1})| \\ &\quad + |(1-u_1)\alpha_1(X_{h2}I_{h2} - X_{h1}I_{h1}) + \alpha_2(X_{h2}I_{b2} - X_{h1}I_{b1})| \\ &\quad + |(1-u_3)\alpha_3(X_{b2}I_{b2} - X_{b1}I_{b1})| \\ &\quad + |(1-u_3)\alpha_3(X_{b2}I_{b2} - X_{b1}I_{b1})| \\ &\le M\left[|X_{h2} - X_{h1}| + |I_{h2} - I_{h1}| + |X_{b2} - X_{b1}| + |I_{b2} - I_{b1}|\right], \end{aligned} \tag{9}$$

where

$$M = \max\left\{2\left[(1-u_1)\alpha_1\frac{\Lambda}{\mu_h} + \alpha_2\frac{\pi}{\mu_b}\right], \right. \\ \left. 2\left[(1-u_3)\alpha_3\frac{\pi}{\mu_b} + \alpha_2\frac{\Lambda}{\mu_h}\right]\right\}. \tag{10}$$

Here the constant M is independent of the state variables X_h, I_h, R_h, X_b, and I_b. Also we get $|(Y_1) - D(Y_2)| \le V|Y_1 - Y_2|$, where $V = \max\{M, \|B\|\}$, which shows that D is uniformly Lipschitz continuous; hence by definition, the solution of the system exists (see [15]).

Let us go back to the optimal control problem (2)–(3). In order to find an optimal solution, first we should find the Lagrangian and Hamiltonian for the optimal control problem. The Lagrangian is

$$L = A_1 I_h + A_2 I_b + \frac{1}{2}\left(C_1 u_1^2 + C_2 u_2^2 + C_3 u_3^2\right). \tag{11}$$

We seek for the minimal value of the Lagrangian. To do this, we define the Hamiltonian H for the control problem, where λ_i, $i = 1, 2, \ldots, 8$, are the adjoint variables:

$$H = A_1 I_h + A_2 I_b + \frac{1}{2}\left(C_1 u_1^2 + C_2 u_2^2 + C_3 u_3^2\right)$$
$$+ \lambda_1 \left[\Lambda - \mu_h X_h - (1 - u_1)\alpha_1 X_h I_h - \alpha_2 X_h I_b + \epsilon_h R_h\right]$$
$$+ \lambda_2 \left[(1 - u_1)\alpha_1 X_h I_h + \alpha_2 X_h I_b - (\mu_h + \phi_h) E_h\right]$$
$$+ \lambda_3 \left[\phi_h E_h - (\rho_h + \beta_h + \mu_h + u_2) I_h\right]$$
$$+ \lambda_4 \left[\rho_h I_h - (\gamma_h + \mu_h) T_h\right]$$
$$+ \lambda_5 \left[\gamma_h T_h + u_2 I_h - (\mu_h + \epsilon_h) R_h\right]$$
$$+ \lambda_6 \left[\pi - \mu_b X_b - (1 - u_3)\alpha_3 X_b I_b\right]$$
$$+ \lambda_7 \left[(1 - u_3)\alpha_3 X_b I_b - (\mu_b + \phi_b) E_b\right]$$
$$+ \lambda_8 \left[\phi_b E_b - (\beta_b + \mu_b) I_b\right]. \tag{12}$$

For the existence of the control system (2), we state and prove the following theorem.

Theorem 1. There exists an optimal control $u* = (u^{*\,1}, u^{*\,2}, u^{*\,3}) \in U$ such that

$$J\left(u_1^*, u_2^*, u_3^*\right) = \min_{(u_1, u_2, u_3) \in U} J\left(u_1, u_2, u_3\right) \tag{13}$$

subject to the control system (2) with the initial conditions (2).

Proof. To prove the existence of an optimal control, we use the result in [16–18]. Note that the control and the state variables are non-negative values. In this minimizing problem, the necessary convexity of the objective functional in u_1, u_2, and u_3 is satisfied. The set of all the control variables $(u_1, u_2, u_3) \in U$ is also convex and closed by definition. The optimal system is bounded which determines the compactness needed for the existence of an optimal control. In addition the integrand in the functional (3), $A_1 I_h + A_2 I_b + (1/2)(C_1 u^2{}_1 + C_2 u^2{}_2 + C_3 u^2{}_3)$, is convex on the control set U. Also we can see that there exist a constant $\rho > 1$ and positive numbers ω_1 and ω_2 such that

$$J\left(u_1, u_2, u_3\right) \geq \omega_1 \left(|u_1|^2 + |u_2|^2 + |u_3|^2\right)^{\rho/2} - \omega_2, \tag{14}$$

because the state variables are bounded, which complete the existence of an optimal control.

In order to derive the necessary conditions, we use Pontryagin's maximum principle [14] as follows. If (x, u) is an optimal solution of an optimal control problem, then there exists a nontrivial vector function $\lambda = (\lambda_1, \lambda_2, ...,)$ satisfying the following equations:

$$\frac{dx}{dt} = \frac{\partial H(t, x, u, \lambda)}{\partial \lambda}, \qquad 0 = \frac{\partial H(t, x, u, \lambda)}{\partial u},$$

$$\frac{d\lambda}{dt} = \frac{\partial H(t, x, u, \lambda)}{\partial x}. \tag{15}$$

We now derive the necessary conditions that optimal control functions and corresponding states must satisfy. In the following theorem, we present the adjoint system and control characterization.

Theorem 2. Let $X_h^*, E_h^*, I_h^*, T_h^*, R_h^*, X_b^*, E_b^*, and\ I_b^*$ be optimal state solutions with associated optimal control variables ($u^*{}_1$, $u^*{}_2$, and $u^*{}_3$) for the optimal control problem (2)–(3). Then there exist adjoint variables λi, for $i = 1, 2, \ldots, 8$, satisfying

$$\frac{d\lambda_1}{dt} = (\lambda_1 - \lambda_2)(1 - u_1)(\alpha_1 I_h + \alpha_2 I_b) + \lambda_1 \mu_h,$$

$$\frac{d\lambda_2}{dt} = (\lambda_2 - \lambda_3)\phi_h + \lambda_2 \mu_h,$$

$$\frac{d\lambda_3}{dt} = (\lambda_1 - \lambda_2)(1 - u_1)\alpha_1 X_h + (\lambda_3 - \lambda_4)\rho_h$$
$$+ \lambda_3(\beta_h + \mu_h + u_2) - \lambda_5 u_2 - A_1,$$

$$\frac{d\lambda_4}{dt} = (\lambda_4 - \lambda_5)\gamma_h + \lambda_4 \mu_h,$$

$$\frac{d\lambda_5}{dt} = (\lambda_5 - \lambda_1)\epsilon_h + \lambda_5 \mu_h,$$

$$\frac{d\lambda_6}{dt} = (\lambda_6 - \lambda_7)(1 - u_3)\alpha_3 I_b + \lambda_6 \mu_b,$$

$$\frac{d\lambda_7}{dt} = (\lambda_7 - \lambda_8)\phi_b + \lambda_7 \mu_b,$$

$$\frac{d\lambda_8}{dt} = (\lambda_6 - \lambda_7)(1 - u_3)\alpha_3 X_b + \lambda_8(\beta_b + \mu_b) - A_2 \tag{16}$$

with transversality conditions

$$\lambda_i\left(t_{end}\right) = 0, \quad i = 1, 2, \ldots, 8. \tag{17}$$

Furthermore the control functions u^*_1, u^*_2, and u^*_3 are given by

$$u_1^* = \max\left\{\min\left\{\frac{(\lambda_2 - \lambda_1)\alpha_1 X_h^* I_h^*}{c_1}, 1\right\}, 0\right\},$$

$$u_2^* = \max\left\{\min\left\{\frac{(\lambda_3 - \lambda_5) I_h^*}{c_2}, 1\right\}, 0\right\},$$

$$u_3^* = \max\left\{\min\left\{\frac{(\lambda_7 - \lambda_6)\alpha_3 X_b^* I_b^*}{c_3}, 1\right\}, 0\right\}. \tag{18}$$

Proof. To determine the adjoint equations and the transversality conditions, we use the Hamiltonian H in (12). The adjoint system results from Pontryagin's maximum principle [14]:

$$\frac{d\lambda_1}{dt} = -\frac{\partial H}{\partial X_h}, \quad \frac{d\lambda_2}{dt} = -\frac{\partial H}{\partial E_h}, \quad \frac{d\lambda_3}{dt} = -\frac{\partial H}{\partial I_h},$$

$$\frac{d\lambda_4}{dt} = -\frac{\partial H}{\partial T_h}, \quad \frac{d\lambda_5}{dt} = -\frac{\partial H}{\partial R_h}, \quad \frac{d\lambda_6}{dt} = -\frac{\partial H}{\partial X_b},$$

$$\frac{d\lambda_7}{dt} = -\frac{\partial H}{\partial E_b}, \quad \frac{d\lambda_8}{dt} = -\frac{\partial H}{\partial I_b} \tag{19}$$

with $\lambda_i(tend)=0$, $i = 1, 2, \ldots, 8$.

To get the characterization of the optimal control given by (18), solving the equations

$$\frac{\partial H}{\partial u_1} = 0, \quad \frac{\partial H}{\partial u_2} = 0, \quad \frac{\partial H}{\partial u_3} = 0, \tag{20}$$

on the interior of the control set, and setting the property of the control space U, we can derive the desired characterization (18).

NUMERICAL RESULTS AND DISCUSSION

In this section, we present a semi-implicit finite difference method by discretizing the interval $[t_0, t_f]$ at the points $ti = t_0 + i_l$, $(i = 0, 1, \ldots, n)$, where l represents the time step such that $t_n = t_f$. We define the state and adjoint variables X_h, E_h, I_h, T_h, R_h, X_b, E_b, I_b, λ_1, λ_2, λ_3, λ_4, λ_5, λ_6, λ_7, and λ_8 and the controls u_1, u_2, and u_3 in terms of nodal points X_h^i, E_h^i I_h^i, T_h^i, R_h^i, X_b^i, E_b^i, I_b^i, λ_1^i, λ_2^i, λ_3^i, λ_4^i, λ_5^i, λ_6^i, λ_7^i, λ_8^i, u_1^i, u_2^i, and u_3^i. By combination of forward and backward difference approximation, the method developed by [19] to adopt it in our case is given by

$$
\begin{aligned}
\frac{X_h^{i+1} - X_h^i}{l} &= \Lambda - \mu_h X_h^{i+1} - \left(1 - u_1^i\right)\alpha_1 X_h^{i+1} I_h^i \\
&\quad - \alpha_2 X_h^{i+1} I_b^i + \epsilon_h R_h^i, \\
\frac{E_h^{i+1} - E_h^i}{l} &= \left(1 - u_1^i\right)\alpha_1 X_h^{i+1} I_h^i + \alpha_2 X_h^{i+1} I_b^i \\
&\quad - \left(\mu_h + \phi_h\right) E_h^{i+1}, \\
\frac{I_h^{i+1} - I_h^i}{l} &= \phi_h E_h^{i+1} - \left(\rho_h + \beta_h + \mu_h + u_2^i\right) I_h^{i+1}, \\
\frac{T_h^{i+1} - T_h^i}{l} &= \rho_h I_h^{i+1} - \left(\gamma_h + \mu_h\right) T_h^{i+1}, \\
\frac{R_h^{i+1} - R_h^i}{l} &= \gamma_h T^{i+1} + u_2^i I_h^{i+1} - \left(\mu_h + \epsilon_h\right) R_h^{i+1}, \\
\frac{X_b^{i+1} - X_b^i}{l} &= \pi - \mu_b X_b^{i+1} - \left(1 - u_3^i\right)\alpha_3 X_b^{i+1} I_b^i, \\
\frac{E_b^{i+1} - E_b^i}{l} &= \left(1 - u_3^i\right)\alpha_3 X_b^{i+1} I_b^i - \left(\mu_b + \phi_b\right) E_b^{i+1}, \\
\frac{I_b^{i+1} - I_b^i}{l} &= \phi_b E_b^{i+1} - \left(\beta_b + \mu_b\right) I_b^{i+1}.
\end{aligned} \tag{21}
$$

To approximate the time derivative of the adjoint variables by the first-ordered backward difference, we use the appropriate scheme as follows:

$$
\begin{aligned}
\frac{\lambda_1^{n-i} - \lambda_1^{n-i-1}}{l} &= \left(\lambda_1^{n-i-1} - \lambda_2^{n-i}\right)\left(1 - u_1^i\right)\left(\alpha_1 I_h^{i+1} + \alpha_2 I_b^{i+1}\right) \\
&\quad + \lambda_1^{n-i-1}\mu_h, \\
\frac{\lambda_2^{n-i} - \lambda_2^{n-i-1}}{l} &= \left(\lambda_2^{n-i-1} - \lambda_3^{n-i}\right)\phi_h + \lambda_2^{n-i-1}\mu_h, \\
\frac{\lambda_3^{n-i} - \lambda_3^{n-i-1}}{l} &= \left(\lambda_1^{n-i-1} - \lambda_2^{n-i-1}\right)\left(1 - u_1^i\right)\alpha_1 X_h^{i+1} \\
&\quad + \lambda_3^{n-i-1}\left(\rho_h + \beta_h + \mu_h + u_2^i\right) \\
&\quad - \lambda_5^{n-i} u_2^i - \lambda_4^{n-i}\rho_h - A_1, \\
\frac{\lambda_4^{n-i} - \lambda_4^{n-i-1}}{l} &= \left(\lambda_4^{n-i-1} - \lambda_5^{n-i}\right)\gamma_h + \lambda_4^{n-i-1}\mu_h, \\
\frac{\lambda_5^{n-i} - \lambda_5^{n-i-1}}{l} &= \left(\lambda_5^{n-i-1} - \lambda_1^{n-i-1}\right)\epsilon_h + \lambda_5^{n-i-1}\mu_h, \\
\frac{\lambda_6^{n-i} - \lambda_6^{n-i-1}}{l} &= \left(\lambda_6^{n-i-1} - \lambda_7^{n-i}\right)\left(1 - u_3^i\right)\alpha_3 I_b^{i+1} + \lambda_6^{n-i-1}\mu_b, \\
\frac{\lambda_7^{n-i} - \lambda_7^{n-i-1}}{l} &= \left(\lambda_7^{n-i-1} - \lambda_8^{n-i}\right)\phi_b + \lambda_7^{n-i-1}\mu_b, \\
\frac{\lambda_8^{n-i} - \lambda_8^{n-i-1}}{l} &= \left(\lambda_6^{n-i-1} - \lambda_7^{n-i-1}\right)\left(1 - u_3^i\right)\alpha_3 X_b^{i+1} \\
&\quad + \lambda_8^{n-i-1}\left(\beta_b + \mu_b\right) - A_2.
\end{aligned} \tag{22}
$$

The algorithm that describes the approximation method for obtaining the optimal control is as follows.

Algorithm 3. Consider the following.

Step 1. Consider $X_h(0) = X_{h0}$, $Eh(0) = E_{h0}$, $Ih(0) = I_{h0}$, $T_h(0) = T_{h0}$, $R_h(0) = R_{h0}$, $X_b(0) = X_{b0}$, $E_b(0) = E_{b0}$, $I_b(0) = I_{b0}$, $u_1(0) = u_2(0) = u_3(0) = 0$, and $\lambda_i(tf)=0$, $i = 1, \ldots, 8$.

Step 2. For $i = 1, \ldots, n-1$, do

$$X_h^{i+1} = \frac{X_h^i + l\left[\Lambda + \epsilon_h R_h^i\right]}{1 + l\left[\mu_h + \left(1 - u_1^i\right)\alpha_1 I_h^i + \alpha_2 I_b^i\right]},$$

$$E_h^{i+1} = \frac{E_h^i + l\left[\left(1 - u_1^i\right)\alpha_1 X_h^{i+1} I_h^i + \alpha_2 X_h^{i+1} I_b^i\right]}{1 + l\left[\mu_h + \phi_h\right]},$$

$$I_h^{i+1} = \frac{I_h^i + l\phi_h E_h^{i+1}}{1 + l\left[\rho_h + \beta_h + \mu_h + u_2^i\right]},$$

$$T_h^{i+1} = \frac{T_h^i + l\rho_h I_h^{i+1}}{1 + l\left[\gamma_h + \mu_h\right]},$$

$$R_h^{i+1} = \frac{R_h^i + l\left[\gamma T_h^{i+1} + u_2^i I_h^{i+1}\right]}{1 + l\left[\mu_h + \epsilon_h\right]},$$

$$X_b^{i+1} = \frac{X_b^i + l\pi}{1 + l\left[\mu_b + \left(1 - u_3^i\right)\alpha_3 I_b^i\right]},$$

$$E_b^{i+1} = \frac{E_b^i + l\left(1 - u_3^i\right)\alpha_3 X_b^{i+1} I_b^i}{1 + l\left[\mu_b + \phi_b\right]},$$

$$I_b^{i+1} = \frac{I_b^i + l\phi_b E_b^{i+1}}{1 + l\left[\beta_b + \mu_b\right]},$$

$$\lambda_1^{n-i-1} = \frac{\lambda_1^{n-i} + l\lambda_2^{n-i}\left(1 - u_1^i\right)\left[\alpha_1 I_h^{i+1} + \alpha_2 I_h^{i+1}\right]}{1 + l\left[\left(1 - u_1^i\right)\alpha_1 I_h^{i+1} + \alpha_2 I_h^{i+1} + \mu_h\right]},$$

$$\lambda_2^{n-i-1} = \frac{\lambda_3^{n-i} + l\lambda_3^{n-i}\phi_h}{1 + l\lambda_2^{n-i-1}\left[\phi_h + \mu_h\right]},$$

$$\lambda_3^{n-i-1} = \lambda_3^{n-i} + l\left[\left(\lambda_2^{n-i-1} - \lambda_1^{n-i-1}\right)\left(1 - u_1^i\right)\alpha_1 X_h^{i+1} + \rho_h\lambda_4^{n-i} + u_2^i\lambda_5^{n-i} + A_1\right] \times \left(1 + l\left[\rho_h + \beta_h + \mu_h + u_2^i\right]\right)^{-1},$$

$$\lambda_4^{n-i-1} = \frac{\lambda_4^{n-i} - l\gamma_h\lambda^{n-i}}{1 - l\left[\gamma_h + \mu_h\right]},$$

$$\lambda_5^{n-i-1} = \frac{\lambda_5^{n-i} + l\epsilon_h\lambda_1^{n-i-1}}{1 + l\left[\epsilon_h + \mu_h\right]},$$

$$\lambda_6^{n-i-1} = \frac{\lambda_6^{n-i} + l\left(1 - u_3^i\right)I_b^{i+1}\lambda_7^{n-i}}{1 + l\left[\mu_b + \left(1 - u_3^i\right)\alpha_3 I_b^{i+1}\right]},$$

$$\lambda_7^{n-i-1} = \frac{\lambda_7^{n-i} + l\phi_b \lambda_8^{n-i}}{1 + l\left[\phi_b + \mu_b\right]},$$

$$\lambda_8^{n-i-1} = \frac{\lambda_8^{n-i} + l\left[\left(\lambda_7^{n-i-1} - \lambda_6^{n-i-1}\right)\left(1 - u_3^i\right)\alpha_3 X_b^{i+1} + A_2\right]}{1 + l\left[\beta_b + \mu_b\right]},$$

$$u_1^{i+1} = \max\left\{\min\left\{\frac{\left(\lambda_2^{n-i-1} - \lambda_1^{n-i-1}\right)\alpha_1 X_h^{i+1} I_h^{i+1}}{c_1}, 1\right\}, 0\right\},$$

$$u_2^{i+1} = \max\left\{\min\left\{\frac{\left(\lambda_3^{n-i-1} - \lambda_5^{n-i-1}\right) I_h^{i+1}}{c_2}, 1\right\}, 0\right\},$$

$$u_3^{i+1} = \max\left\{\min\left\{\frac{\left(\lambda_7^{n-i-1} - \lambda_6^{n-i-1}\right)\alpha_3 X_b^{i+1} I_b^{i+1}}{c_3}, 1\right\}, 0\right\}, \tag{23}$$

end for.

Step 3. For $i = 1, \ldots, n-1$, write $X_h^*(t_i) = X_h^i$, $E_h^*(t_i) = E_h^i$, $I_h^*(t_i) = I_h^i$, $T_h^*(t_i) = T_h^i$, $R_h^*(t_i) = R_h^i$, $X_b^*(t_i) = X_b^i$, $E_b^*(t_i) = E_b^i$, $I_b^*(t_i) = I_b^i$, $u_1^*(t_i) = u_1^i$, $u_2^*(t_i) = u_2^i$, and $u_3^*(t_i) = u_3^i$ end for.

We have plotted susceptible, exposed, infected, treated, and recovered population of humans and susceptible, exposed, and infected population of birds with and without control by considering parameter values given in Table 1, with initial values $X_h(0) = 40$, $E_h(0) = 20$, $I_h(0) = 10$, $T_h(0) = 8$, $R_h(0) = 5$, $X_b(0) = 70$, $E_b(0) = 30$, and $I_b(0) = 20$. In each of the given figures the undashed and the dashed lines represent the individuals without and with control, respectively. The weight constant of the objective functional $A_1 = 0.035$, $A_2 = 0.0015$, $C_1 = 0.05$, $C_2 = 0.08$, and $C_3 = 0.002$ is considered for numerical simulation.

Figure 1 shows that the population of the susceptible humans increased after control, in Figures 2 and 3 we see that the number of the exposed and infected humans with control decreased more sharply than that without control, Figure 4 shows that per day clinically reported humans decreased after control, Figure 5 shows that the number of recovered human individuals increased after control, Figure 6 shows that the birds population increased after control, and in Figures 7 and 8 we see that the number of the exposed and infected birds decreased after control. Figure 9 shows the graphs of control variables.

Table 1. Parameter values used for numerical simulations.

Notation	Parameters definition	Value
Λ	Recruitment rate of humans	1.7/day
α_1	Effective contact rate between X_h and I_h	0.3
α_2	Effective contact rate between X_h and I_b	0.15
α_3	Effective contact rate between X_b and I_b	0.12
π	Recruitment rate of birds	5/day
ϵ_h	Rate of immunity loss	0.005
μ_h	Natural death rate of humans	0.1/day
μ_b	Natural death rate of birds	0.3/day
γ_h	Recovery due to treatment	0.3
ρ_h	Treatment rate	0.75
β_h	Disease induced death rate in humans	0.4/day
β_b	Disease induced death rate in birds	3.7/day
ϕ_h	Progression rate from E_h to I_h	0.45
ϕ_b	Progression rate from E_b to I_b	0.85

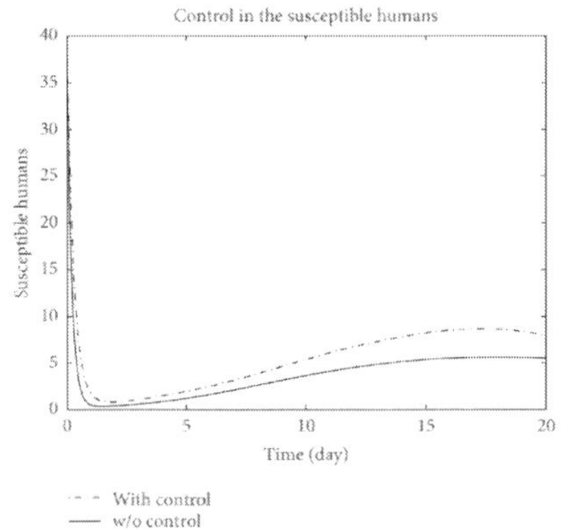

Figure 1. The plot represents the population of susceptible humans with and without control.

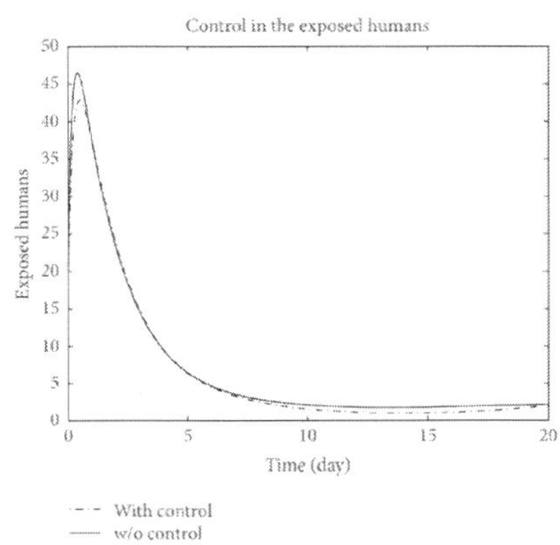

Figure 2. The plot represents the population of exposed humans with and without control.

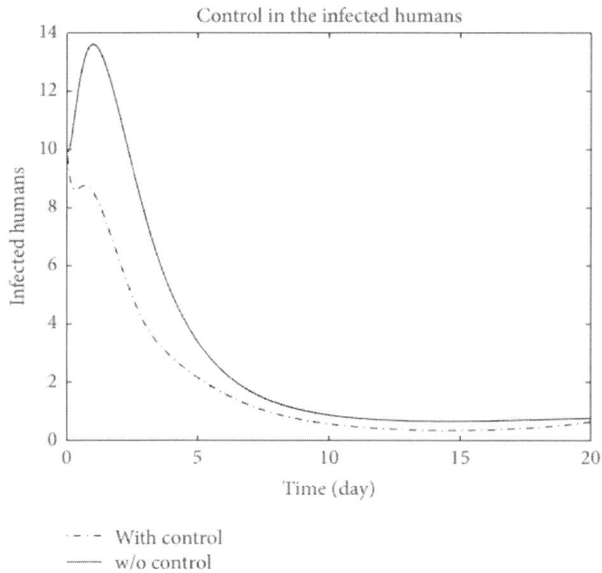

Figure 3. The plot represents the population of infected humans with and without control.

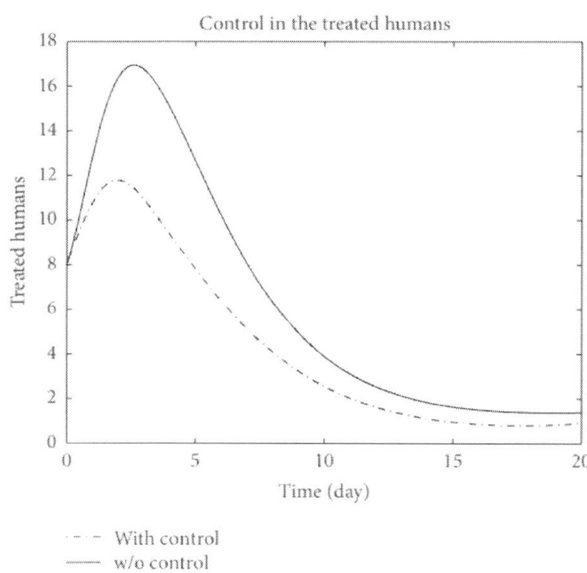

Figure 4. The plot represents the population of treated humans with and without control.

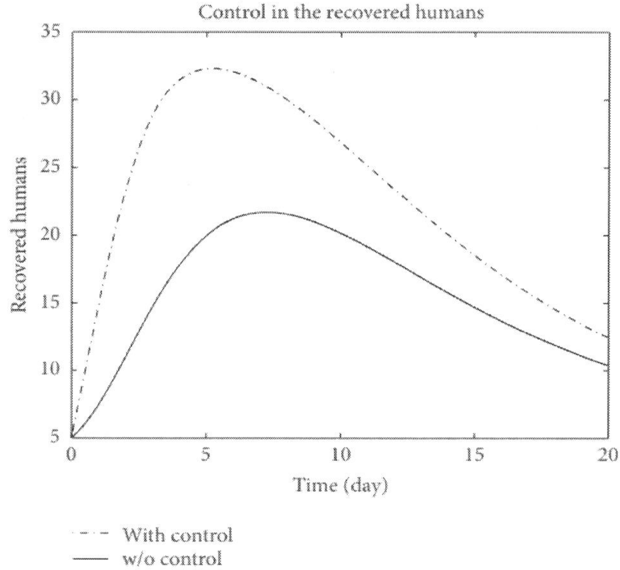

Figure 5. The plot represents the population of recovered humans with and without control.

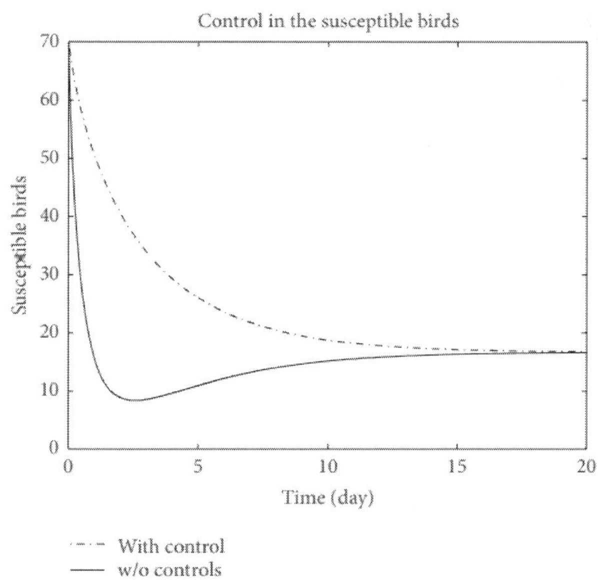

Figure 6. The plot represents the population of susceptible birds with and without control.

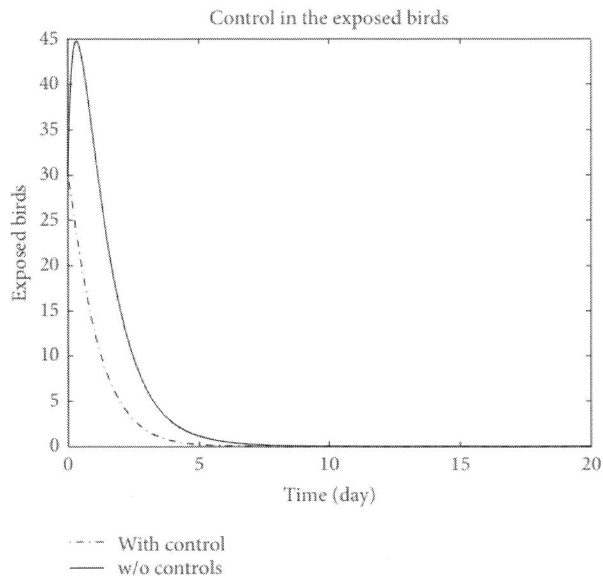

Figure 7. The plot represents the population of exposed birds with and without control.

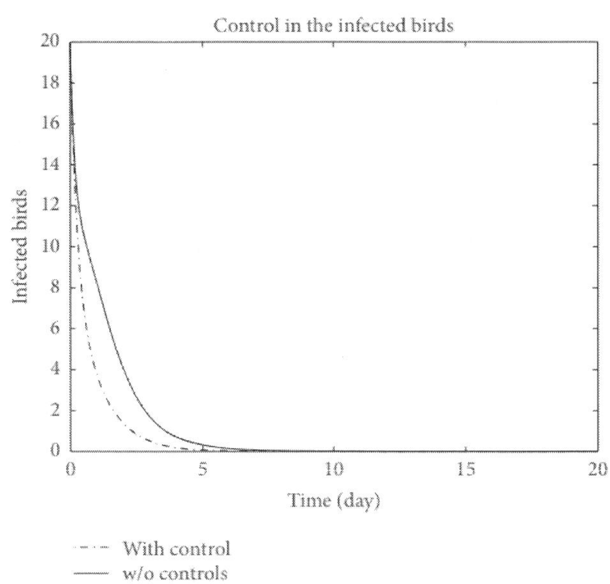

Figure 8. The plot represents the population of infected birds with and without control.

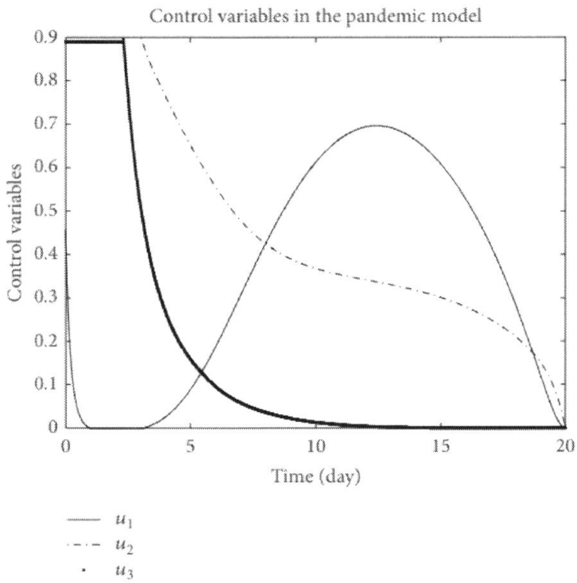

Figure 9. The plot represents the control variables u_1, u_2, and u_3

CONCLUSION

An optimal control problem of the transmission dynamics of the human-avian influenza disease has been presented. We sought to determine three types of control functions associated with isolation and antiviral treatment of the clinically infectious individuals and elimination of the infected birds. Our control plots indicated that the number of exposed, infected, and hospitalized humans and the number of exposed, infected birds decreased in the optimal system. We developed the necessary conditions for the existence of the optimal control by using Pontryagin's maximum principle. Using the state and adjoint system together with the characterization of the optimal control, we solved the problem numerically by an efficient numerical method based on optimal control with the estimated parameter values based on avian influenza. The results show that the control practices are very effective in reducing the incidence of infectious population.

CONFLICT OF INTERESTS

The authors declare that there is no conflict of interests regarding the publication of this paper.

REFERENCES

1. Centers for Disease Control and Prevention, "Key Facts about Avian Influenza (Bird Flu) and Avian Influenza A, (H5N1) Virus," 2006, http://www.cdc.gov/flu/avian/gen-info/facts.htm.
2. M. Martcheva, B. M. Bolker, and R. D. Holt, "Vaccine-induced pathogen strain replacement: what are the mechanisms?" *Journal of the Royal Society Interface*, vol. 5, no. 18, pp. 3–13, 2008.
3. Y. Yang, M. E. Halloran, J. D. Sugimoto, and I. M. Longini Jr., "Detecting human-to-human transmission of avian influenza A (H5N1)," *Emerging Infectious Diseases*, vol. 13, no. 9, pp. 1348–1353, 2007.
4. M. Zaman, S. Ashraf, N. A. Dreyer, and S. Toovey, "Human infection with avian influenza virus, Pakistan, 2007," *Emerging Infectious Diseases*, vol. 17, no. 6, pp. 1056–1059, 2011.
5. World Health Organization Europe, "Making preparation count: lessons from the avian influenza outbreak in Turkey," 2006, http://www.euro.who.int/__data/assets/pdf_file/0018/90513/E89139.pdf.
6. D. Butler, "Family tragedy spotlights flu mutations," *Nature*, vol. 442, no. 7099, pp. 114–115, 2006.
7. N. M. Ferguson, D. A. T. Cummings, S. Cauchemez et al., "Strategies for containing an emerging influenza pandemic in southeast Asia," *Nature*, vol. 437, no. 7056, pp. 209–214, 2005.
8. R. F. Grais, J. H. Ellis, and G. E. Glass, "Assessing the impact of airline travel on the geographic spread of pandemic influenza," *European Journal of Epidemiology*, vol. 18, no. 11, pp. 1065–1072, 2003
9. I. M. Longini Jr., A. Nizam, S. Xu et al., "Containing pandemic influenza at the source," *Science*, vol. 309, no. 5737, pp. 1083–1087, 2005.
10. S. Iwami, Y. Takeuchi, and X. Liu, "Avian-human influenza epidemic model," *Mathematical Biosciences*, vol. 207, no. 1, pp. 1–25, 2007.
11. S. Lee, G. Chowell, and C. Castillo-Chávez, "Optimal control for pandemic influenza: the role of limited antiviral treatment and isolation," *Journal of Theoretical Biology*, vol. 265, no. 2, pp. 136–150, 2010.
12. R. Ullah, G. Zaman, and S. Islam, "Prevention of influenza pandemic by multiple control strategies," *Journal of Applied Mathematics*, vol. 2012, Article ID 294275, 14 pages, 2012.

13. R. Ullah, G. Zaman, S. Islam, and M. R. Muhyuddin, "Global dynamics of the avian-human influenza with horizontal transmission in human population," *Life Science Journal*, vol. 9, no. 4, pp. 5747–5753, 2012.
14. L. S. Pontryagin, V. G. Boltyanskii, R. V. Gamkrelidze, and E. F. Mishchenko, *The Mathematical Theory of Optimal Processes*, John Wiley & Sons, Hoboken, NJ, USA, 1962.
15. G. Birkhoff and G. C. Rotta, *Ordinary Differential Equations*, John Wiley & Sons, New York, NY, USA, 4th edition, 1989.
16. A. A. Lashari and G. Zaman, "Optimal control of a vector borne disease with horizontal transmission," *Nonlinear Analysis: Real World Applications*, vol. 13, no. 1, pp. 203–212, 2012
17. D. L. Lukes, *Differntial Equations: Classical to Controlled*, vol. 162 of *Mathematics in Science and Engineering*, Academic Press, New York, NY, USA, 1982.
18. G. Zaman, Y. Han Kang, and I. H. Jung, "Stability analysis and optimal vaccination of an SIR epidemic model," *BioSystems*, vol. 93, no. 3, pp. 240–249, 2008.
19. A. B. Gumel, P. N. Shivakumar, and B. M. Sahai, "A mathematical model for the dynamics of HIV-1 during the typical course of infection," in *Proceedings of the 3rd World Congress of Non-linear Analysts*, vol. 47, pp. 2073–2083, 2001.

9

Prevalence, Severity and Mortality Associated with COPD and Smoking in patients with COVID-19: A Rapid Systematic Review and Meta-Analysis

Jaber S. Alqahtani[1,2], Tope Oyelade[3] , Abdulelah M. Aldhahir[1,4,] Saeed M. Alghamdi[5,6], Mater Almehmadi[7] , Abdullah S. Alqahtani[2] , Shumonta Quaderi[1] , Swapna Mandal[1,8], John R. Hurst[1]

[1] UCL Respiratory, University College London, London, United Kingdom,

[2] Department of Respiratory Care, Prince Sultan Military College of Health Sciences, Dammam, Saudi Arabia,

[3] UCL Institute for Liver and Digestive Health, London, United Kingdom,

[4] Respiratory Care Department, Faculty of Applied Medical Sciences, Jazan University, Jazan, Saudi Arabia,

[5] National Heart and Lung Institute, Imperial College London, London, United Kingdom,

[6] Faculty of Applied Medical Sciences, Umm Al-Qura University, Makkah, Saudi Arabia,

[7] UCL Institute for Risk and Disaster, London, United Kingdom,

[8] Royal Free London NHS Foundation Trust, London, United Kingdom

Citation: Prevalence, Severity and Mortality associated with COPD and Smoking in patients with COVID-19: A Rapid Systematic Review and Meta-Analysis. Alqahtani JS, Oyelade T, Aldhahir AM, Alghamdi SM, Almehmadi M, Alqahtani AS, et al. (2020). PLoS ONE 15(5): e0233147. https://doi.org/10.1371/journal.pone.0233147

ABSTRACT

Background

Coronavirus disease 2019 (COVID-19) is an evolving infectious disease that dramatically spread all over the world in the early part of 2020. No studies have yet summarized the potential severity and mortality risks caused by COVID-19 in patients with chronic obstructive pulmonary disease (COPD), and we update information in smokers.

Methods

We systematically searched electronic databases from inception to March 24, 2020. Data were extracted by two independent authors in accordance with the Preferred Reporting Items for Systematic Reviews and Meta-Analyses guidelines. Study quality was assessed using a modified version of the Newcastle-Ottawa Scale. We synthesized a narrative from eligible studies and conducted a meta-analysis using a random-effects model to calculate pooled prevalence rates and 95% confidence intervals (95%CI).

Results

In total, 123 abstracts were screened and 61 full-text manuscripts were reviewed. A total of 15 studies met the inclusion criteria, which included a total of 2473 confirmed COVID-19 patients. All studies were included in the meta-analysis. The crude case fatality rate of COVID-19 was 7.4%. The pooled prevalence rates of COPD patients and smokers in COVID-19 cases were 2% (95% CI, 1%–3%) and 9% (95% CI, 4%–14%) respectively. COPD patients were at a higher risk of more severe disease (risk of severity = 63%, (22/35) compared to patients without COPD 33.4% (409/1224) [calculated RR, 1.88 (95% CI, 1.4–2.4)]. This was associated with higher mortality (60%). Our results showed that 22% (31/139) of current smokers and 46% (13/28) of ex-smokers had severe complications. The calculated RR showed that current smokers were 1.45 times more likely [95% CI: 1.03–2.04] to have severe complications compared to former and never smokers. Current smokers also had a higher mortality rate of 38.5%.

Conclusion

Although COPD prevalence in COVID-19 cases was low in current reports, COVID-19 infection was associated with substantial severity and mortality rates in COPD. Compared to former and never smokers, current smokers were at greater risk of severe complications and higher mortality rate. Effective preventive measures are required to reduce COVID-19 risk in COPD patients and current smokers.

IMPACT

In this systematic review and meta-analysis of 15 studies including a total of 2473 confirmed COVID-19 patients, COVID-19 was associated with greater severity and mortality in COPD patients and current smokers. Despite the low prevalence of COPD and smoking in COVID-19 cases, this result emphasizes attention to effective preventive measures to support reducing the burden of COVID-19 in these vulnerable populations. This has the potential to improve outcomes for patients and lessen the burden on health services.

INTRODUCTION

The emergence in Wuhan City, Hubei Province of China of a novel pneumonia of unknown origin on the 31st of December, 2019 was the start of an outbreak which would later be declared a pandemic by the World Health Organization (WHO) [1]. The name COVID-19 (acronym for "coronavirus disease 2019") was coined on the 11th of February 2020 to describe presentation with severe acute respiratory disease [2]. COVID-19 is caused by a novel strain of coronavirus, the severe acute respiratory syndrome coronavirus 2 (SARS-CoV-2). This virus belongs to the family of single stranded RNA viruses some of which have been previously described to be responsible for the Severe Acute Respiratory Syndrome (SARS) and Middle East Respiratory Syndrome (MERS) [3, 4]. Although the symptoms and clinical presentation of COVID-19 is similar to SARS and MERS, the rate of spread is greater [5]. As of 25th of March 2020, the total number of confirmed case of COVID-19 stands at 459,419 with 20,818 deaths globally and the number is projected to increase [6].

Chronic Obstructive Pulmonary Disease (COPD) is a common, persistent and preventable dysfunction of the lung associated with limitation in airflow. COPD is a complex disease associated with abnormalities of the airway and/ or alveoli which is predominantly caused by exposure to noxious gases and particulates over a long period [7]. With a global prevalence of 251 million cases in 2016, and 3.17 million (5%) deaths in 2015 alone, COPD was ranked by the WHO as the third leading cause of death especially with a particular burden in low- and middle-income countries [8, 9]. This burden is predicted to grow due mainly to increased global exposure to tobacco, aging populations, poor awareness and inadequate access to diagnosis [10]. COPD exacerbations are a major event in the natural history of the disease associated with worsening of symptoms often resulting in hospitalization and poor prognosis [11]. Various factors have been described to contribute to acute worsening of COPD, however, viral infection remains the main trigger, including seasonal coronaviruses [12–14].

Since the emergence of COVID-19, scientists around the world continue to better understand the clinical, diagnostic and prognostic characteristic of the disease. While over 150 papers, editorials and comments have been written about COVID-19 as of 24th of March 2020, there is none dedicated to the specific risk posed to patients with a previous history of COPD. This review addresses this gap in knowledge with the aim of assisting clinicians to assess the prognosis of COVID-19 infection in patients with COPD and those with smoking history.

Methods

This systematic review was conducted in accordance with the Preferred Reporting in Systematic Reviews and Meta-Analyses (PRISMA) guidelines [15]. We prospectively registered this review to Prospero (registration number is: CRD42020175518).

We searched MEDLINE and Google scholar from inception date to March 16, 2020. The search was updated on March 24, 2020. We used an extensive search strategy developed by a specilaised librarian for retrieving this type of evidence, which included the reference list of eligible papers and published pre-print papers. All retrieved studies were exported into EndNote to remove duplicates. The remaining studies were exported to Rayyan software for title, abstracts and full text screening by two independent reviewers.

Inclusion and exclusion criteria

Eligible studies were those that investigated: epidiomolgical, clinical charcterstics and features of COVID-19 and prevelance of chronic diseases specifically COPD in their analysis. We excluded the following: studies that did not report COPD, studies that included respiratory diseases but did not specifically analyze COPD, only children, editorials, correspondence letters, reviews, qualitative studies, theses, non-English language and non-full text articles.

Data collection

Two authors (JSA and TO) independently screened titles and abstracts of potential studies and conflicts were resolved through discussion between the two. Full-text articles of potential studies were then independently read by two authors (JSA and TO) to identify studies meeting the inclusion criteria. The reference lists from all identified studies and reviews were scrutinized for eligible articles. Disagreement on selected papers was resolved through discussion with a third author (AA).

Quality assessment

Two authers independently evaluated the methodological quality of included studies using a modified version of the Newcastle-Ottawa Scale (NOS) [16]. It includes seven domains, each one of these domains was scored from 0 (high risk of bias) to 3 (low risk of bias) and we took a mean of the domains to result in a score between 0 and 3, where a higher score represents a lower risk of bias. Any disagreement in the quality assessment was resolved by discussion with a third author.

Data synthesis

We completed meta-analysis to calculate the pooled prevalence of COPD and current smokers among those patients with confrmed COVID-19.

The output was generated using the Stata procedure Metaprop. Owing to heterogeneity within and between studies, we used the random-effects model in Stata/SE 15. Data were displayed using forest plots. We examined between-study heterogeneity using the I^2 statistic.

A narrative synthesis of the results was conducted considering the prevalence, disease severity and mortality among COVID-19 COPD patients and smoking status. We defined COVID-19 severity as those who were admitted to intensive care unit (ICU), had severe, oxygenation, needed mechanical ventilation or death. Since our meta-analysis was performed to estimate the pooled prevalence and not assessing treatment effects, the assessment for potential risk of bias that could affect the results was not indicated.

Pre-print

As this is a rapidly evolving area, we also searched pre-print literature and whilst not incorporating such data in the formal analysis, provide narrative synthesis where such pre-print data are relevant to our main findings.

RESULTS

An initial search generated 123 potentially relevant papers, of which two were immediately excluded due to duplication. After the first screening of title and abstracts, 61 papers were potentially relevant according to the inclusion criteria. An additional 46 papers were excluded after full-text review, which resulted in 15 studies that satisfied all criteria. The reference list of the relevant papers was also examined (see Fig 1, PRISMA flow diagram).

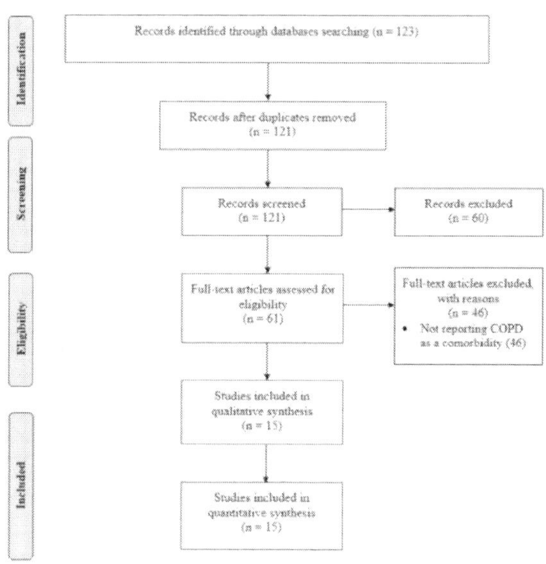

Figure 1. PRISMA flow chart for the included studies.

Description of included studies

A summary of the included studies is presented in Table 1, which included a total of 2473 confirmed COVID-19 patients. Of those patients, only 58 (2.3%) had COPD as a comorbidity. The sample size of the included studies ranged from 21 to 1099 patients.

Table 1. Characteristics of the included studies.

Study Name	Country/ Type of study	Sample size (M = male, F = female)	Age mean ±SD or median, range	Current Smokers	COPD patients	COPD Survived	COPD Non-Survived	Mortality rate for study/ COPD	Severe cases for study/ severe COPD Cases
Arentz M et al. 2020 [17]	United states/ Washington Retrospective study from February 20, 2020, and March 5, 2020,	21 M = 11 (52%) F = 10 (48%)	70 (43–59)	NA	7 (33.3%)	NA	NA	11 (52.4%)	NA
Guan et el. 2020 [18]	China/ 30 provinces Retrospective study from December 11, 2019, and January 29, 2020	1099 M = 640 (58.3%) F = 459 (41 7%)	47 (35–58) Severe 52 (40–65) Non-severe 45 (34–57)	137 (12.5%) Severe cases = 29/172 (16.9%)	12 (1.1%)	NA	NA	15 (1.4%)	173 (15.7%)/ 6 (3.4%)
Huang Y et al. 2020 [19]	China /Hubei province Retrospective study from December 2019 to January 2020	34 M = 14 (41.2% F = 20 (58.8%)	56.24 ± 17.14	NA	3 (8.82%)	NA	NA	NA	NA
Huang et al. 2020 [20]	China/Wuhan Prospective study from Dec 16, 2019, to Jan 2, 2020	41 M = 30 (73%) F = 11 (27%)	49 (41–58·)	3 (7%)	1 (2%)	NA	NA	6 (15%)	29 (70.7%)/ 1 (3.44%)

Liu et al. 2020 [21]	China/Hubei province Retrospective study from December 30, 2019 to January 24, 2020	137 M = 61 (44.5%) F = 76 (55.5%)	57 (20–83)	NA	2 (1.5%)	NA	NA	16 (11.7%)	NA
Mo et al. 2020 [22]	China/Wuhan retrospective study from January 1st to February 5th	155 M = 86 (55.5%) F = 69 (45.5%)	54 (42–66)	6 (3.9%)	5 (3.2%)	NA	NA	NA	85 (54.8%)/ 4 (3.7%)
Wang et al. 2020 [23]	China/Wuhan Retrospective, single-center case series from January 1 to January 28, 2020	138 M = 75 (54.3%) F = 63 (45.7%)	56 (42–68) ICU (n = 36) = 66 (57–78) Non-ICU (n = 102) = 51 (37–62)	NA	4 (2.9%)	NA	NA	6 (4.3%)	3 (8.3%)
Wu et al. 2020 [24]	China/Wuhan Retrospective study from December 25, 2019, and January 26, 2020	201 M = 128 (63.7%) F = 73 (36.3%)	51 (43–60)	NA	5 (2.5%)	NA	NA	44 (21.9%)	NA
Wu J et al. 2020 [25]	China/ outside of the Wuhan area) Prospective study from January to February 2020	80 M = 42 (52%) F = 38 (48%)	44 (11)	26 (33%)	3 (4%)	NA	NA	NA	NA
Xu X et al. 2020 [26]	China/Guangzhou retrospective study from January 23, 2020, and February 4, 2020	90 M = 39 (43%) F = 51 (57%)	50 (18–86)	NA	1 (1%)	NA	NA	NA	NA
Xu XW et al. 2020 [27]	China/Zhejiang province Retrospective study from 10 January 2020 to 26 January 2020	62 M = 35 (56%) F = 27 (44%)	41 (32–52)	NA	1 (2%)	NA	NA	0	45 (72.5%)/ 1 (3%)

Yang et al. 2020 [28]	China/Wuhan Retrospective study from late December, 2019, and Jan 26, 2020.	52 M = 35 (67%) F = 17 (33%)	59.7 (13.3)	2 (4%) Survived = 2 (10%)	4 (8%)	2 (10%)	2 (6%)	32 (61.5%)/2 (50%)	NA
Zhang et al. 2020 [29]	China/Wuhan Retrospective study from January 16 to February 3, 2020	140 M = 71 (50.7%) F = 69 (49.3%)	57 (25–87)	2 (1.4%) Severe cases = 2 (3.4)	2 (1.4)	NA	NA	NA	58 (41.4%)/ 2 (3.4%)
Zhou et al. 2020 [30]	China/Wuhan retrospective cohort study from Dec 29, 2019 to Jan 31, 2020	191 M = 119 (62%) F = 72 (38%)	56 (46-–67)	11 (6%) Non-survivor = 5 (9%) Survivor = 6 (4%)	6 (3%)	2 (1%)	4 (7%)	54 (28.3%)/4 (66.7%)	NA
Zhu et al. 2020 [31]	China/ Anhui province Retrospective study from 24 January 2020 and 20 February 2020	32 M = 15 (47%) F = 17 (53%)	46(35–52)	6 (18.7%)	2 (6%)	NA	NA	NA	NA

Most of the studies were retrospectively conducted and all were studied in China, except one in the United States. The risk of bias ranged from 0.4 to 2.7; nine studies scored ≥ 2, which indicates low risk of bias. The overall crude mortality rate for COVID-19 was 7.4% (184/2473). There were no reports of the baseline characteristics of the 58 COPD patients that described their age, or severity of COPD airflow limitation. Smoking status was reported in eight studies, in which 193 confirmed COVID-19 cases were current smokers.

Prevalence of COPD in confirmed COVID-19 cases

The pooled prevalence of COVID-19 patients who had COPD from all studies was 2% (95% CI, 1%–3%), Fig 2. The overall I^2 was 41.90%, p 0.04.

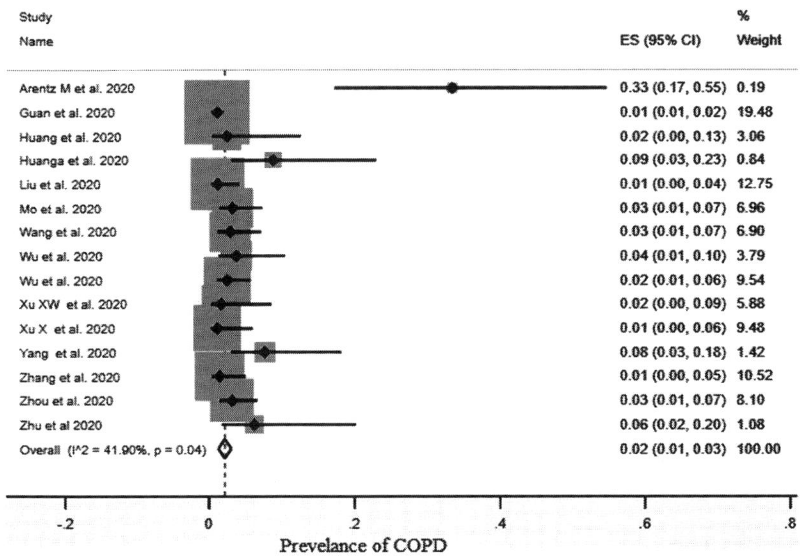

Figure 2. Pooled prevalence of COPD with confirmed COVID-19, (ES: Effect size).

Disease severity and mortality among COVID-19 COPD patients

Seven studies that included 35 COPD patients reported COVID severity in their analysis. With 63% (22/35) patient reported as severe compared to 37% (13/35) non-severe, this shows that COPD patients are at a higher risk of more severe COVID-2019 compared to patients without COPD 33.4% (409/1224) [calculated RR, 1.88 (95% CI, 1.4–2.4)]. Data from two studies [28, 30] including COPD patients with confirmed COVID-19 demonstrate 60% (6/10) mortality rate compared to mortality rate in patients without COPD 55% (86/157), [calculated RR, 1.10 (95% CI, 0.6–1.8)].

Smoking history and risk of COVID-2019

Smoking exposure including (current and ex-smokers) was reported in eight studies, with 221 confirmed COVID-19 cases. We assessed the prevalence of current smokers in all studies using meta-analysis of proportions of current smokers that had confirmed COVID-19. There was a pooled prevalence of 9%, (95% CI, 4%–14%), Fig 3. The overall I^2 was 92.49%, p 0.00. Indeed, 22.30% (31/139) of current smokers and 46% (13/28) of ex-smokers had severe complications. Calculating the RR from these two studies showed that current smokers (31/108) were 1.45 times more likely [RR = 1.45, 95% CI: 1.03–2.04] to have severe complications compared to former and never smokers (147/808) patients [18, 29]. A higher mortality rate of 38.5% (5/13) was also seen among current smokers.

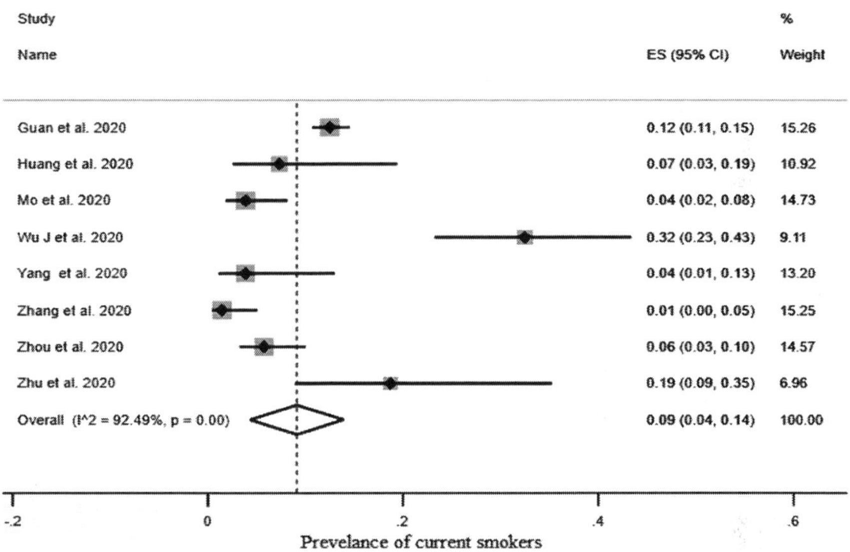

Figure 3. Pooled prevalence of current smokers with confirmed COVID-19, (ES: Effect size).

DISCUSSION

To the best of our knowledge, this is the first systematic review and meta-analysis to develop an informed understanding of the prevalence, severity and mortality of COPD patients diagnosed with COVID-19. We provide an updated report in relation to smokers [32]. Our main outcomes show that the prevalence of COPD in COVID-19 patients was low, but that the risk of severity (63%) and mortality (60%) were high, which indicates COPD patients with confirmed COVID-19 are at a greater risk of severe complications and death. Furthermore, the prevalence of current smokers in COVID-19 patients was 9% (95% CI, 4%–14%), and this was also associated with greater severity (22.30%) and mortality (38.5%).

We report a low prevalence of COPD patients in COVID-19 case series compared to the latest COPD prevalence rate in China, which was 13.6% (95% CI 12·0–15·2) and the global prevalence of COPD (9–10%) [33, 34]. There was a low heterogeneity among the studies, which was 41.90%, p 0.04. This is expected since clinical and methodological diversity always occur in a meta-analysis. We speculate that patients may have not been diagnosed. Having a reliable estimate of the prevalence of COPD in COVID-19 cases, and likely outcomes, is crucial to ensure specific successful global preventive and treatment strategies for COPD patients. Bearing this in mind, in the included studies there was no report on COPD severity data and COPD-related comorbidities, which prevents us from assessing the impact of such essential information.

Although the COPD prevalence was not high in the included confirmed COVID-19 cases, COVID-19 causes a substantial burden on COPD patients with increased disease severity. This summary agrees with other results currently available only in pre-print [35, 36]. Viral infections in COPD patients increase systemic inflammation with slow recovery of reported symptoms [37, 38]. In addition to the effects of COVID-19, patients with COPD have various comorbidities, some of which are associated with increased risk of hospitalization [39–41]. According to recent systematic reviews, the prevalence of comorbidities in COVID-19 patients was high and these comorbidities were associated with increased disease severity [42, 43]. Most of the studies that reported COPD severity defined severe cases as those who were admitted to intensive care unit (ICU), had severe oxygenation, needed mechanical ventilation or death [18, 19, 22, 23, 27, 29]. These studies did not give details about the underlying COPD. It would have been interesting to know if the frequency of previous exacerbations was linked to increased complications in COPD patients diagnosed with COVID-19. In general, those with severe case of COVID-19 were older and had more coexisting comorbidities than those with mild illness.

Data from two studies [28, 30] describing COPD patients with confirmed COVID-19 show a higher mortality rate at 60%. This concurs with pre-print studies that found similarly high mortality rates [35, 44–47]. Despite the small number of patients that were analyzed, this increases concerns about the prognosis of this vulnerable population. However, this high mortality rate could be attributed to several factors. The majority of COPD patients have various comorbidities that may also be associated with mortality and related conditions may have been underreported because of the difficulties finding the specific cause of mortality [48]. Moreover, in patients with severe COPD, respiratory failure is the principal cause of mortality and this demands ICU intervention. It is possible that limited access to respiratory support as part of COVID-19 management may be contributing to this mortality, dependent on critical care capacity in each hospital or region. According to a recent COVID-19 report from Italy, the surge in patients requiring intensive care has been unmanageable, with 12% of positive cases requiring ICU admission [49], more than that reported in China [18, 28]. As a consequence, patients were dying because mechanical ventilation could not be offered, on top of acute shortage of clinicians who were able to manage those patients [50, 51]. Until now, the particular mechanisms of how COVID-19 increases COPD severity and mortality is unknown, and undoubtedly more research is needed to find the possible mechanisms that linked COVD-19 and increased severity and mortality of COPD patients.

Concerning smoking and COVID-19, our data showed a pooled prevalence of 9% current smokers, (95% CI, 4%–14%), lower than the reported prevalence of smoking in China that was 25.2% (25.1–25.4) [52]. The overall I^2 was 92.49% among the studies, which indicates considerable heterogeneity. However, this was address by using random effect model to incorporate between-study variability of effect sizes. Such heterogeneity is expected due to different study designs, regions and outcomes. Interestingly, we found that 22.30% (31/139) of current smokers and 46% (13/28) of ex-smokers had severe complications associated and greater mortality reaching 38.5% in current smokers. The calculated RR from two studies showed that current smokers were 1.45 times more likely [RR = 1.45, 95% CI: 1.03–2.04] to have severe complications compared to former and never smokers [18, 29].

The impact of smoking history on vulnerability to COVID-19 has been explored but there is limited data on the contribution of tobacco smoking to the spread of and poor outcome in COVID-19. A recent systematic review on COVID-19 and smoking including five studies found that smoking was most likely associated with the negative outcomes. This recommended further research to explore this in more detail due to limited studies [32]. Evidence from other respiratory viruses, respiratory syncytial virus, has shown that inhaled tobacco smoke raises the transmission rate and severity of viral respiratory tract infections [53]. It seems there is underlying mechanisms behind this prevalence, as smoking has been related to higher expression of

ACE2 (the receptor for SARS-CoV-2) (pre-print) [54]. However, as more reports globally from diverse racial and genetic contexts become available, differences in the production of ACE2 can be further evaluated and linked to how they lead to COVID-19 vulnerability in different groups [55].

To our knowledge, the present study is the first to systematically evaluate existing literature with a focus on risk of COVID-19 on COPD. For the first time, we conducted a meta-analysis using a random effects model to calculate the pooled prevalence of COPD in confirmed COVID-19 and examined outcomes. This increased the generalizability of our findings, as heterogeneity was addressed by incorporating between-study variability of effect sizes. We showed that COPD and smoking in COVID-19 is associated with greater disease severity and higher mortality. This review has some limitations. Few studies were eligible for inclusion and most of them come from China. Second, heterogeneity exists in location, setting, and design. The reported clinical characteristics were not available in most of the studies at the time of analyses. This work has a number of clinical and research implications. It highlights the global prevalence and the clinical effects of COVID-19 on COPD patients and smokers. As COPD patients are at an increased risk of severe outcomes if they became infected with COVID-19, it is recommended that patients and clinicians establish effective plans for ensuring prevention, such as using tele-medicine to ensure that COPD receive the best care [56, 57]. We strongly advocate public awareness campaigns concentrating on ways to achieve smoking cessation among smokers, and it is possible that an improvement in cessation rates will help to reduce the spread of SARS-CoV-2. Future studies should investigate the mechanisms between COPD, smoking and COVID-19 infection.

CONCLUSION

Though COPD prevalence in reported COVID-19 cases is low, COVID-19 infection is associated with significant severity and mortality in COPD. There was also increased risk of severe disease and mortality in current smokers. Effective preventive measures are urgently required to reduce COVID-19 risk on COPD patients and current smokers.

ACKNOWLEDGMENTS

We thank Jennifer Ford, Evidence Services Librarian at Royal Free Hospital Medical Library, UCL Library Services, London UK for her assistance and support in refining the search strategy.

FUNDING STATEMENT

The author(s) received no specific funding for this work.

REFERENCES

1. WHO. Coronavirus disease (COVID-2019) Situation Report 2020 [Available from: https://www.who.int/emergencies/diseases/novel-coronavirus-2019/events-as-they-happen.
2. WHO. Situation Report -22 2020 [Available from: https://www.who.int/docs/default-source/coronaviruse/situation-reports/20200211-sitrep-22-ncov.pdf?sfvrsn=fb6d49b1_2.
3. Su S, Wong G, Shi W, Liu J, Lai ACK, Zhou J, et al. Epidemiology, Genetic Recombination, and Pathogenesis of Coronaviruses. Trends Microbiol. 2016;24(6):490–502. 10.1016/j.tim.2016.03.003
4. Zhu N, Zhang D, Wang W, Li X, Yang B, Song J, et al. A Novel Coronavirus from Patients with Pneumonia in China, 2019. N Engl J Med. 2020;382(8):727–33. 10.1056/NEJMoa2001017
5. Peeri NC, Shrestha N, Rahman MS, Zaki R, Tan Z, Bibi S, et al. The SARS, MERS and novel coronavirus (COVID-19) epidemics, the newest and biggest global health threats: what lessons have we learned? Int J Epidemiol. 2020.
6. Zhu H, Wang L, Fang C, Peng S, Zhang L, Chang G, et al. Clinical analysis of 10 neonates born to mothers with 2019-nCoV pneumonia. Translational Pediatrics.9(1):51–60. 10.21037/tp.2020.02.06
7. GOLD. [Available from: https://goldcopd.org/wp-content/uploads/2019/11/GOLD-2020-REPORT-ver1.0wms.pdf.
8. Organization WH. Chronic obstructive pulmonary disease (COPD) 2020 [Available from: https://www.who.int/news-room/fact-sheets/detail/chronic-obstructive-pulmonary-disease-(copd)
9. De Ramón Fernández A, Ruiz Fernández D, Marcos-Jorquera D, Gilart Iglesias V. Support System for Early Diagnosis of Chronic Obstructive Pulmonary Disease Based on the Service-Oriented Architecture Paradigm and Business Process Management Strategy: Development and Usability Survey Among Patients and Health Care Providers. J Med Internet Res. 2020;22(3):e17161 10.2196/17161
10. Miravitlles M, Soriano JB, Garcia-Rio F, Munoz L, Duran-Tauleria E, Sanchez G, et al. Prevalence of COPD in Spain: impact of undiagnosed COPD on quality of life and daily life activities. Thorax. 2009;64(10):863–8. 10.1136/thx.2009.115725
11. Seemungal TA, Donaldson GC, Paul EA, Bestall JC, Jeffries DJ,

Wedzicha JA. Effect of exacerbation on quality of life in patients with chronic obstructive pulmonary disease. Am J Respir Crit Care Med. 1998;157(5 Pt 1):1418–22.

12. White AJ, Gompertz S, Stockley RA. Chronic obstructive pulmonary disease. 6: The aetiology of exacerbations of chronic obstructive pulmonary disease. Thorax. 2003;58(1):73–80. 10.1136/thorax.58.1.73
13. Papi A, Bellettato CM, Braccioni F, Romagnoli M, Casolari P, Caramori G, et al. Infections and airway inflammation in chronic obstructive pulmonary disease severe exacerbations. Am J Respir Crit Care Med. 2006;173(10):1114–21. 10.1164/rccm.200506-859OC
14. Wedzicha JA, Seemungal TA. COPD exacerbations: defining their cause and prevention. Lancet (London, England). 2007;370(9589):786–96.
15. Moher D, Shamseer L, Clarke M, Ghersi D, Liberati A, Petticrew M, et al. Preferred reporting items for systematic review and meta-analysis protocols (PRISMA-P) 2015 statement. Syst Rev. 2015;4:1 10.1186/2046-4053-4-1
16. GA Wells BS, D O'Connell, J Peterson, V Welch, M Losos, P Tugwell. The Newcastle-Ottawa Scale (NOS) for assessing the quality of nonrandomised studies in meta-analyses [cited 2019 23/01]. Available from: http://www.ohri.ca/programs/clinical_epidemiology/oxford.asp
17. Arentz M, Yim E, Klaff L, Lokhandwala S, Riedo FX, Chong M, et al. Characteristics and Outcomes of 21 Critically Ill Patients With COVID-19 in Washington State. JAMA. 2020.
18. Guan WJ, Ni ZY, Hu Y, Liang WH, Ou CQ, He JX, et al. Clinical Characteristics of Coronavirus Disease 2019 in China. New England Journal of Medicine. 2020;28:28.
19. Huang Y, Tu M, Wang S, Chen S, Zhou W, Chen D, et al. Clinical characteristics of laboratory confirmed positive cases of SARS-CoV-2 infection in Wuhan, China: A retrospective single center analysis. Travel Medicine & Infectious Disease.101606.
20. Huang C, Wang Y, Li X, Ren L, Zhao J, Hu Y, et al. Clinical features of patients infected with 2019 novel coronavirus in Wuhan, China. Lancet (London, England).395(10223):497–506.
21. Liu K, Fang YY, Deng Y, Liu W, Wang MF, Ma JP, et al. Clinical characteristics of novel coronavirus cases in tertiary hospitals in Hubei Province. Chinese Medical Journal. 2020;07:07.
22. Mo P, Xing Y, Xiao Y, Deng L, Zhao Q, Wang H, et al. Clinical

characteristics of refractory COVID-19 pneumonia in Wuhan, China. Clinical Infectious Diseases. 2020;16:16.

23. Wang D, Hu B, Hu C, Zhu F, Liu X, Zhang J, et al. Clinical Characteristics of 138 Hospitalized Patients With 2019 Novel Coronavirus-Infected Pneumonia in Wuhan, China. Jama. 2020;07:07.
24. Wu C, Chen X, Cai Y, Xia J, Zhou X, Xu S, et al. Risk Factors Associated With Acute Respiratory Distress Syndrome and Death in Patients With Coronavirus Disease 2019 Pneumonia in Wuhan, China. JAMA Internal Medicine. 2020;13:13.
25. Wu J, Wu X, Zeng W, Guo D, Fang Z, Chen L, et al. Chest CT Findings in Patients with Corona Virus Disease 2019 and its Relationship with Clinical Features. Investigative Radiology. 2020;21:21.
26. Xu X, Yu C, Qu J, Zhang L, Jiang S, Huang D, et al. Imaging and clinical features of patients with 2019 novel coronavirus SARS-CoV-2. European Journal of Nuclear Medicine & Molecular Imaging. 2020;28:28.
27. Xu XW, Wu XX, Jiang XG, Xu KJ, Ying LJ, Ma CL, et al. Clinical findings in a group of patients infected with the 2019 novel coronavirus (SARS-Cov-2) outside of Wuhan, China: retrospective case series. BMJ.368:m606 10.1136/bmj.m606
28. Yang X, Yu Y, Xu J, Shu H, Xia J, Liu H, et al. Clinical course and outcomes of critically ill patients with SARS-CoV-2 pneumonia in Wuhan, China: a single-centered, retrospective, observational study. The Lancet Respiratory medicine. 2020.
29. Zhang JJ, Dong X, Cao YY, Yuan YD, Yang YB, Yan YQ, et al. Clinical characteristics of 140 patients infected with SARS-CoV-2 in Wuhan, China. Allergy. 2020;19:19.
30. Zhou F, Yu T, Du R, Fan G, Liu Y, Liu Z, et al. Clinical course and risk factors for mortality of adult inpatients with COVID-19 in Wuhan, China: a retrospective cohort study. Lancet (London, England). 2020;11:11.
31. Zhu W, Xie K, Lu H, Xu L, Zhou S, Fang S. Initial clinical features of suspected Coronavirus Disease 2019 in two emergency departments outside of Hubei, China. Journal of Medical Virology. 2020;13:13.
32. Vardavas CI, Nikitara K. COVID-19 and smoking: A systematic review of the evidence. Tob Induc Dis. 2020;18(March).
33. Fang L, Gao P, Bao H, Tang X, Wang B, Feng Y, et al. Chronic

obstructive pulmonary disease in China: a nationwide prevalence study. The Lancet Respiratory medicine. 2018;6(6):421–30. 10.1016/S2213-2600(18)30103-6

34. Halbert RJ, Natoli JL, Gano A, Badamgarav E, Buist AS, Mannino DM. Global burden of COPD: systematic review and meta-analysis. The European respiratory journal. 2006;28(3):523–32. 10.1183/09031936.06.00124605
35. Guan W-j, Liang W-h, Zhao Y, Liang H-r, Chen Z-s, Li Y-m, et al. Comorbidity and its impact on 1,590 patients with COVID-19 in China: A Nationwide Analysis. medRxiv. 2020.
36. Zhang G, Hu C, Luo L, Fang F, Chen Y, Li J, et al. Clinical features and outcomes of 221 patients with COVID-19 in Wuhan, China. medRxiv. 2020.
37. Aaron SD, Donaldson GC, Whitmore GA, Hurst JR, Ramsay T, Wedzicha JA. Time course and pattern of COPD exacerbation onset. Thorax. 2012;67(3):238–43. 10.1136/thoraxjnl-2011-200768
38. Williams NP, Ostridge K, Devaster J-M, Kim V, Coombs NA, Bourne S, et al. Impact of radiologically stratified exacerbations: insights into pneumonia aetiology in COPD. Respir Res. 2018;19(1):143-. 10.1186/s12931-018-0842-8
39. Terzano C, Colamesta V, Unim B, Romani S, Meneghini A, Volpe G, et al. Chronic obstructive pulmonary disease (COPD) exacerbation: impact of comorbidities on length and costs during hospitalization. European review for medical and pharmacological sciences. 2017;21(16):3680–9.
40. Hurst JR, Skolnik N, Hansen GJ, Anzueto A, Donaldson GC, Dransfield MT, et al. Understanding the impact of chronic obstructive pulmonary disease exacerbations on patient health and quality of life. European journal of internal medicine. 2020;73:1–6. 10.1016/j.ejim.2019.12.014
41. Aldhahir AM, Rajeh AMA, Aldabayan YS, Drammeh S, Subbu V, Alqahtani JS, et al. Nutritional supplementation during pulmonary rehabilitation in COPD: A systematic review. Chron Respir Dis. 2020;17:1479973120904953-.
42. Yang J, Zheng Y, Gou X, Pu K, Chen Z, Guo Q, et al. Prevalence of comorbidities in the novel Wuhan coronavirus (COVID-19) infection: a systematic review and meta-analysis. International journal of infectious diseases: IJID: official publication of the International

Society for Infectious Diseases. 2020.

43. Rodriguez-Morales AJ, Cardona-Ospina JA, Gutierrez-Ocampo E, Villamizar-Pena R, Holguin-Rivera Y, Escalera-Antezana JP, et al. Clinical, laboratory and imaging features of COVID-19: A systematic review and meta-analysis. Travel medicine and infectious disease. 2020:101623.
44. Zhang B, Zhou X, Qiu Y, Feng F, Feng J, Jia Y, et al. Clinical characteristics of 82 death cases with COVID-19. medRxiv. 2020.
45. Shi Q, Zhao K, Yu J, Feng J, Zhao K, Zhang X, et al. Clinical characteristics of 101 non-surviving hospitalized patients with COVID-19: A single center, retrospective study. medRxiv. 2020.
46. Huang Y, Zhou H, Yang R, Xu Y, Feng X, Gong P. Clinical characteristics of 36 non-survivors with COVID-19 in Wuhan, China. medRxiv. 2020.
47. Li X, Wang L, Yan S, Yang F, Xiang L, Zhu J, et al. Clinical characteristics of 25 death cases infected with COVID-19 pneumonia: a retrospective review of medical records in a single medical center, Wuhan, China. medRxiv. 2020.
48. Sin DD, Anthonisen NR, Soriano JB, Agusti AG. Mortality in COPD: role of comorbidities. European Respiratory Journal. 2006;28(6):1245 10.1183/09031936.00133805
49. Grasselli G, Pesenti A, Cecconi M. Critical Care Utilization for the COVID-19 Outbreak in Lombardy, Italy: Early Experience and Forecast During an Emergency Response. JAMA. 2020.
50. Paterlini M. On the front lines of coronavirus: the Italian response to covid-19. Bmj. 2020;368:m1065 10.1136/bmj.m1065
51. Rosenbaum L. Facing Covid-19 in Italy—Ethics, Logistics, and Therapeutics on the Epidemic's Front Line. New England Journal of Medicine. 2020.
52. Wang M, Luo X, Xu S, Liu W, Ding F, Zhang X, et al. Trends in smoking prevalence and implication for chronic diseases in China: serial national cross-sectional surveys from 2003 to 2013. The Lancet Respiratory Medicine. 2019;7(1):35–45. 10.1016/S2213-2600(18)30432-6
53. Groskreutz DJ, Monick MM, Babor EC, Nyunoya T, Varga SM, Look DC, et al. Cigarette smoke alters respiratory syncytial virus-induced apoptosis and replication. American journal of respiratory cell and molecular biology. 2009;41(2):189–98. 10.1165/rcmb.2008-0131OC
54. Cai G, Cui X, Zhu X, Zhou J. A Hint on the COVID-19 Risk: Population

Disparities in Gene Expression of Three Receptors of SARS-CoV. 2020.

55. Cai H. Sex difference and smoking predisposition in patients with COVID-19. The Lancet Respiratory medicine. 2020.
56. Hollander JE, Carr BG. Virtually Perfect? Telemedicine for Covid-19. New England Journal of Medicine. 2020.
57. Alrajeh AM, Aldabayan YS, Aldhair AM, Pickett E, Quaderi SA, Alqahtani JS, et al. Global use, utility, and methods of tele-health in COPD: a health care provider survey. Int J Chron Obstruct Pulmon Dis. 2019;14:1713–9. 10.2147/COPD.S202640

10

Impact of Self-Imposed Prevention Measures and Short-Term Government-Imposed Social Distancing on Mitigating and Delaying a COVID-19 Epidemic: A Modelling Study

Alexandra Teslya[1,] Thi Mui Pham[1], Noortje G. Godijk, ,[1] Mirjam E. Kretzschmar[1], Martin C. J. Bootsma,[1,2] and Ganna Rozhnova[3]

[1] Julius Center for Health Sciences and Primary Care, University Medical Center Utrecht, Utrecht University, Utrecht, The Netherlands

[2] Mathematical Institute, Utrecht University, Utrecht, The Netherlands

[3] BioISI—Biosystems & Integrative Sciences Institute, Faculdade de Ciências, Universidade de Lisboa, Lisboa, Portugal

ABSTRACT

Background

The coronavirus disease (COVID-19) caused by the severe acute respiratory syndrome coronavirus 2 (SARS-CoV-2) has spread to nearly every country

Citation: Impact of self-imposed prevention measures and short-term government-imposed social distancing on mitigating and delaying a COVID-19 epidemic: A modelling study. Teslya A, Pham TM, Godijk NG, Kretzschmar ME, Bootsma MCJ, Rozhnova G (2020). PLoS Med 17(7): e1003166. https://doi.org/10.1371/journal. pmed.1003166

in the world since it first emerged in China in December 2019. Many countries have implemented social distancing as a measure to "flatten the curve" of the ongoing epidemics. Evaluation of the impact of government-imposed social distancing and of other measures to control further spread of COVID-19 is urgent, especially because of the large societal and economic impact of the former. The aim of this study was to compare the individual and combined effectiveness of self-imposed prevention measures and of short-term government-imposed social distancing in mitigating, delaying, or preventing a COVID-19 epidemic.

Methods and findings

We developed a deterministic compartmental transmission model of SARS-CoV-2 in a population stratified by disease status (susceptible, exposed, infectious with mild or severe disease, diagnosed, and recovered) and disease awareness status (aware and unaware) due to the spread of COVID-19. Self-imposed measures were assumed to be taken by disease-aware individuals and included hand washing, mask-wearing, and social distancing. Government-imposed social distancing reduced the contact rate of individuals irrespective of their disease or awareness status. The model was parameterized using current best estimates of key epidemiological parameters from COVID-19 clinical studies. The model outcomes included the peak number of diagnoses, attack rate, and time until the peak number of diagnoses. For fast awareness spread in the population, self-imposed measures can significantly reduce the attack rate and diminish and postpone the peak number of diagnoses. We estimate that a large epidemic can be prevented if the efficacy of these measures exceeds 50%. For slow awareness spread, self-imposed measures reduce the peak number of diagnoses and attack rate but do not affect the timing of the peak. Early implementation of short-term government-imposed social distancing alone is estimated to delay (by at most 7 months for a 3-month intervention) but not to reduce the peak. The delay can be even longer and the height of the peak can be additionally reduced if this intervention is combined with self-imposed measures that are continued after government-imposed social distancing has been lifted. Our analyses are limited in that they do not account for stochasticity, demographics, heterogeneities in contact patterns or mixing, spatial effects, imperfect isolation of individuals with severe disease, and reinfection with COVID-19.

Conclusions

Our results suggest that information dissemination about COVID-19, which causes individual adoption of hand washing, mask-wearing, and social distancing, can be an effective strategy to mitigate and delay the epidemic. Early initiated short-term government-imposed social distancing can buy time for healthcare systems to prepare for an increasing COVID-19 burden. We stress the importance of disease awareness in controlling the ongoing epidemic and recommend that, in addition to policies on social distancing, governments and public health institutions mobilize people to adopt self-imposed measures with proven efficacy in order to successfully tackle COVID-19.

AUTHOR SUMMARY

Why was this study done?

- As of May 2020, the coronavirus disease (COVID-19) caused by the novel coronavirus (SARS-CoV-2) has spread to nearly every country in the world since it first emerged in China in December 2019.
- Confronted with a COVID-19 epidemic, public health policy makers in different countries are seeking recommendations on how to delay and/or flatten its peak.
- Evaluation of the impact of social distancing mandated by the governments in many countries and of other prevention measures to control further spread of COVID-19 is urgent, especially because of the large societal and economic impact of the former.

What did the researchers do and find?

- We developed a transmission model to evaluate the impact of self-imposed measures (hand washing, mask-wearing, and social distancing) due to awareness of COVID-19 and of short-term government-imposed social distancing on the epidemic dynamics.
- We showed that self-imposed measures can prevent a large epidemic if their efficacy exceeds 50%.
- We estimate that short-term government-imposed social distancing that is initiated early into the epidemic can buy time (at

most 7 months for a 3-month intervention) for healthcare systems to prepare for an increasing COVID-19 burden.

- The delay to the peak number of diagnoses can be even longer and the height of the peak can be additionally reduced if the same intervention is combined with self-imposed measures that are continued after lifting government-imposed social distancing.

What do these findings mean?

- Raising awareness of self-imposed measures such as hand washing and mask-wearing is crucial in controlling the ongoing epidemic.
- Short-term early initiated government-imposed social distancing combined with self-imposed measures provides essential time for increasing capacity of healthcare systems and can significantly mitigate the epidemic.
- In addition to policies on social distancing, governments and public health institutions should continuously mobilize people to adopt self-imposed measures with proven efficacy in order to successfully tackle COVID-19.

INTRODUCTION

As of May 5, 2020, the novel coronavirus (SARS-CoV-2) has spread worldwide and only 13 countries have not reported any cases. It has caused over 3,640,835 confirmed cases of COVID-19 and nearly 255,100 deaths since the detection of its outbreak in China on December 31, 2019 [1]. On March 11, the World Health Organization officially declared the COVID-19 outbreak a pandemic [1]. Several approaches aimed at the containment of SARS-CoV-2 in China were unsuccessful. Airport screening of travelers was hampered by a potentially large number of asymptomatic cases and the possibility of presymptomatic transmission [2–4]. Quarantine of 14 days combined with fever surveillance was insufficient in containing the virus due to the high variability of the incubation period [5].

Now that SARS-CoV-2 has extended its range of transmission in all parts of the world, it is evident that many countries face a large COVID-19 epidemic [6]. Initial policies regarding COVID-19 prevention were mainly limited to reporting cases, strict isolation of severe symptomatic cases, home isolation of mild cases, and contact tracing [7]. However, due to the potentially high contribution of asymptomatic and presymptomatic spread [8], these case-based interventions are likely insufficient in containing a COVID-19 epidemic unless they are highly effective [8–11]. Given the rapid rise in cases and the risk of exceeding critical care bed capacities, many countries have implemented social distancing as a short-term measure aiming at reducing the contact rate in the population and, subsequently, transmission [6, 12]. Several governments have imposed nationwide partial or complete lockdowns by closing schools, public places, and nonessential businesses, canceling mass events, and issuing stay-at-home orders [6]. Previous studies on the 1918 influenza pandemic showed that such mandated interventions were effective in reducing transmission, but their timing and magnitude had a profound influence on the course of the epidemic [13–18]. These short-term interventions were associated with a high risk of epidemic resurgence and their impact was limited if introduced too late or lifted too early [13–16].

Self-imposed prevention measures such as hand washing, mask-wearing, and social distancing could also contribute to slowing down the epidemic [19, 20]. Alcohol-based sanitizers are effective in removing the SARS coronavirus from hands [21], and hand washing with soap may have a positive effect on reducing the transmission of respiratory infections [22]. Surgical masks, often worn for their perceived protection, are not designed nor certified to protect against respiratory hazards, but they can stop droplets being spread from infectious individuals [23–25]. Information dissemination and official recommendations about COVID-19 can create awareness and motivate individuals to adopt such measures. Previous studies emphasized the importance of disease awareness for changing the course of an epidemic [26–28].

Depending on the rate and mechanism of awareness spread, the awareness process can reduce the attack rate of an epidemic or prevent it completely [26], but it can also lead to undesirable outcomes such as the appearance of multiple epidemic peaks [27, 28]. The secondary epidemic waves may appear as the result of individuals relaxing adherence to self-imposed measures prematurely in a population where the susceptible pool following the first wave is still significantly large and disease has not been completely eliminated. It is essential to assess under which conditions the spread of disease awareness that instigates self-imposed measures can be a viable strategy for COVID-19 control.

The comparison of the effectiveness of early implemented short-term government-imposed social distancing and self-imposed prevention measures on reducing the transmission of SARS-CoV-2 are currently missing but are of crucial importance in the attempt to stop its spread. If a COVID-19 epidemic cannot be prevented, it is important to know how to effectively diminish and postpone the epidemic peak to give healthcare professionals more time to prepare and react effectively to an increasing healthcare burden. Moreover, given that several countries have peaked in cases, the importance of evaluating the effect of self-imposed measures after lifting lockdown measures is profound.

Using a transmission model, we evaluated the impact of self-imposed measures (hand washing, mask-wearing, and social distancing) due to awareness of COVID-19 and of a short-term government-imposed social distancing intervention on the peak number of diagnoses, attack rate, and time until the peak number of diagnoses since the first case. We provide a comparative analysis of these interventions as well as of their combinations and assess the range of intervention efficacies for which a COVID-19 epidemic can be mitigated, delayed, or even prevented completely. Qualitatively, these results will aid public health professionals to compare and select a combination of interventions for designing effective outbreak control policies.

METHODS

Baseline transmission model

We developed a deterministic compartmental model describing SARS-CoV-2 transmission in a population stratified by disease status (Fig 1). In this baseline model, individuals are classified as susceptible (S), latently infected (E), infectious with mild disease (I_M), infectious with severe disease (I_S), diagnosed and isolated (I_D), and recovered after mild or severe disease (R_M and R_S, respectively). Susceptible individuals (S) can become latently infected (E) through contact with infectious individuals (I_M and I_S), with the force of infection dependent on the fractions of the population in I_M and I_S compartments. A proportion of the latently infected individuals (E) will go to the I_M compartment, and the remaining E individuals will go to the I_S compartment. We assume that infectious individuals with mild disease (I_M) do not require medical attention and recover (R_M) without being conscious of having contracted COVID-19. Infectious individuals with severe disease (I_S) are unable to recover without medical help, and subsequently get diagnosed and isolated (I_D) (in, e.g., hospitals, long-term care facilities, nursing homes) and know or suspect they have COVID-19 when they are detected. Therefore, the diagnosed compartment I_D contains infectious individuals with severe disease who are both officially diagnosed and get treatment in healthcare institutions and those who are not officially diagnosed but have disease severe enough to suspect they have COVID-19 and require isolation. For simplicity, isolation of these individuals is assumed to be perfect until recovery (R_S), and hence they neither contribute to transmission nor to the contact process. Given the timescale of the epidemic and the lack of reliable reports on reinfections, we assume that recovered individuals (R_M and R_S) cannot be reinfected. The infectivity of infectious individuals with mild disease is lower than the infectivity of infectious individuals with severe disease [29]. Natural birth and death processes are neglected, as the timescale of the epidemic is short compared to the mean life span of individuals. However, isolated infectious individuals with severe disease (I_D) may be removed from the population due to disease-associated mortality.

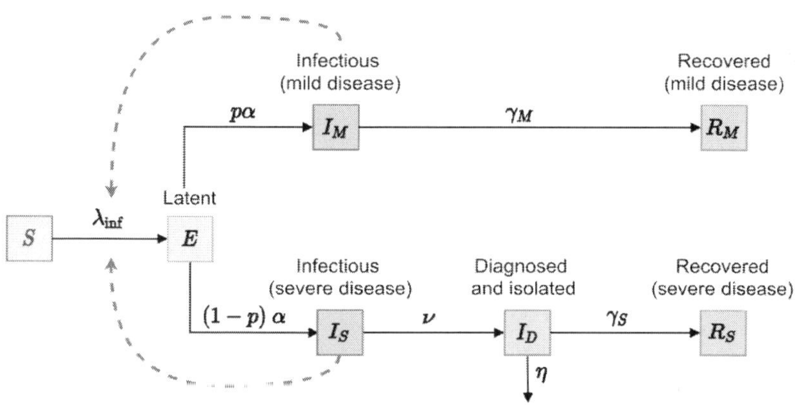

Figure 1. Schematic of the baseline transmission model.

Black arrows show epidemiological transitions. Red dashed arrows indicate the compartments contributing to the force of infection. Susceptible persons (S) become latently infected (E) with the force of infection $_{inf}$ via contact with infectious individuals in two infectious classes (I_M and I_S). Individuals leave the E compartment at rate α. A proportion p of the latently infected individuals (E) will go to the I_M compartment, and the proportion ($1-p$) of E individuals will go to the I_S compartment. Infectious individuals with mild disease (I_M) recover without being conscious of having contracted COVID-19 (R_M) at rate γ_M. Infectious individuals with severe disease (I_S)

are diagnosed and kept in isolation (I_D) at rate ν until they recover (R_S) at rate γ_S or die at rate η. Table 1 provides the description and values of all parameters.

Transmission model with disease awareness

In the extended model with disease awareness, the population is stratified not only by the disease status but also by the awareness status into disease-aware (S^a, E^a, IaM, IaS, IaD, and RaM) and disease-unaware (S, E, I_M, I_S, I_D, and R_M) (Fig 2A). Disease awareness is a state that can be acquired as well as lost. Disease-aware individuals are distinguished from unaware individuals in two essential ways. First, infectious individuals with severe disease who are disease-aware (IaS) get diagnosed and isolated faster (IaD), stay in isolation for a shorter period of time, and have lower disease-associated mortality than the same category of unaware individuals. The assumption we make here is that disease-aware individuals (IaS) recognize they may have COVID-19 on average faster than disease-unaware individuals (I_S) and get medical help earlier, which leads to a better prognosis of IaD individuals as compared to I_D individuals. Second, disease-aware individuals are assumed to use self-imposed measures such as hand washing, mask-wearing, and self-imposed social distancing that can lower their susceptibility, infectivity, and/or contact rate. Individuals who know or suspect their disease status (I_D, IaD, and R_S) do not adapt any such measures because they assume that they cannot contract the disease again. Hence, they are excluded from the awareness transition process and their behavior in the contact process is identical to disease-unaware individuals.

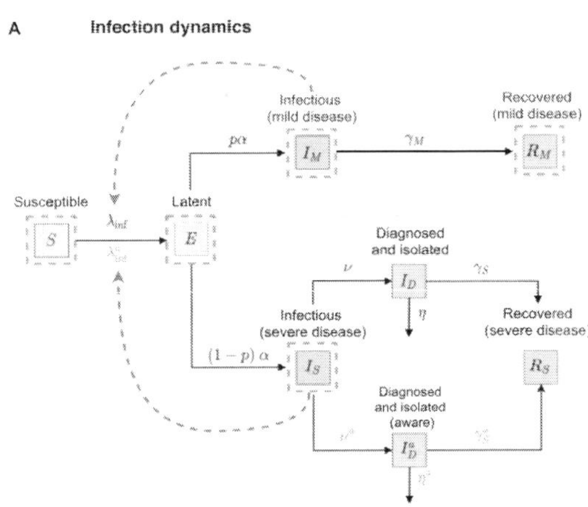

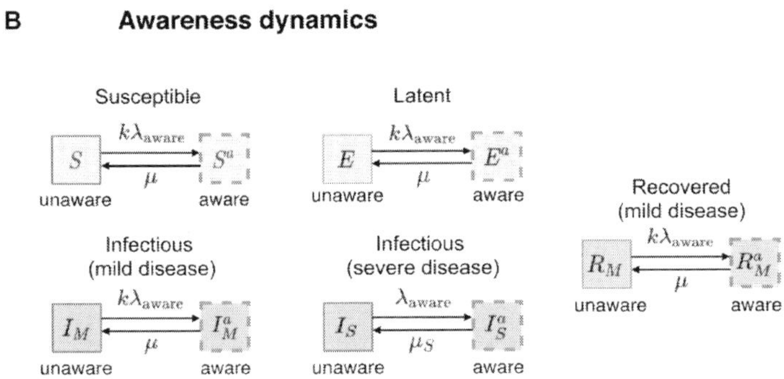

Figure 2. Schematic of the transmission model with disease awareness.

(A) Shows epidemiological transitions in the transmission model with awareness (black arrows). The orange dashed lines indicate the compartments that participate in the awareness dynamics. The red dashed arrows indicate the compartments contributing to the force of infection. Disease-aware susceptible individuals (S^a) become latently infected (E^a) through contact with infectious individuals (I_M, I_S, IaM, and IaS) with the force of infection λainf. Infectious individuals with severe disease who are disease-aware (IaS) get diagnosed and isolated (IaD) at rate ν^a, recover at rate γaS, and die from disease at rate η^a. (B) Shows awareness dynamics. Infectious individuals with severe disease (I_S) acquire disease awareness (IaS) at rate $_{\text{aware}}$ proportional to the rate of awareness spread and to the current number of diagnosed individuals (I_D and IaD) in the population. As awareness fades, these individuals return to the unaware state at rate μ_S. The acquisition rate of awareness (k_{aware}) and the rate of awareness fading (μ) are the same for individuals of types S, E, I_M, and R_M, where k is the reduction in susceptibility to the awareness acquisition compared to I_S individuals. Table 1 provides the description and values of all parameters.

Similarly to Perra and colleagues [27], disease-unaware individuals acquire disease awareness at a rate proportional to the rate of awareness spread and to the current number of diagnosed individuals (I_D and IaD) in the population (Fig 2B). We assume that awareness fades and individuals return to the unaware state at a constant rate. The latter means that they no longer use self-imposed measures. For simplicity, we assume that awareness acquisition and fading rates are the same for individuals of type S, E, I_M, and R_M. However, the rate of awareness acquisition is faster and the fading rate is slower for infectious individuals with severe disease (I_S) than for the remaining disease-aware population.

Prevention measures

We considered short-term government intervention aimed at fostering social distancing in the population and a suite of measures that may be self-imposed by disease-aware individuals, i.e., mask-wearing, hand washing, and self-imposed social distancing.

Mask-wearing

Mask-wearing, while often adapted as a protective measure, may be ineffective in reducing the individual's susceptibility because laypersons, i.e., not medical professionals, are unfamiliar with correct procedures for its use (e.g., often engage in face-touching and mask adjustment) [36]. However, mask-wearing reduces infectious output [25], and therefore we assume that this measure lowers only the infectivity of disease-aware infectious individuals (IaM and IaS) with an efficacy ranging from 0% (zero efficacy) to 100% (full efficacy).

Hand washing

Because infectious individuals may transmit the virus to others without direct physical contact, we assume that hand washing only reduces one's susceptibility. The efficacy of hand washing is described by the reduction in susceptibility (i.e., probability of transmission per single contact) of susceptible disease-aware individuals (S^a), which ranges from 0% (zero efficacy) to 100% (full efficacy). Because transmission can possibly occur through routes other than physical contact, hand washing may not provide 100% protection to those who practice it.

Self-imposed social distancing

Disease-aware individuals who consider themselves susceptible may also practice social distancing, i.e., maintaining distance to others and avoiding congregate settings. As a consequence, this measure leads to a change in mixing patterns in the population. The efficacy of social distancing of disease-aware individuals is described by the reduction in their contact rate, which is varied from 0% (no social distancing or zero efficacy) to 100% (complete self-isolation or full efficacy). Because contacts might not be eliminated entirely (e.g., household contacts remain), realistic values of the efficacy of self-imposed social distancing can be close to but may never reach 100%.

Short-term government-imposed social distancing

Governments may decide to promote social distancing policies through interventions such as school and workplace closures or by issuing stay-at-home orders and bans on large gatherings.

These lockdown policies will cause a community-wide contact rate reduction, regardless of the awareness status. Here, we assume that the government-imposed social distancing is initiated if the number of diagnosed individuals exceeds a certain threshold (10–1,000 persons) and terminates after a fixed period of time (1–3 months).

As such, the intervention is implemented early into the epidemic. Government-imposed social distancing may be partial or complete depending on its efficacy, i.e., the reduction of the average contact rate in the population, which ranges from 0% (no distancing) to 100% (complete lockdown). Because during a lockdown, some contacts in the population cannot be eliminated (e.g., household contacts), realistic values of the efficacy of government-imposed social distancing can be close to but never reach 100%. For example, a 73% reduction in the average daily number of contacts was observed during the lockdown in the United Kingdom [37], but the reduction could be different in countries with more or less stringent lockdown.

Model output

The model outputs are the peak number of diagnoses, attack rate (a proportion of the population that recovered or died after severe infection), the time to the peak number of diagnoses since the first case, and the probability of infection during the course of an epidemic.

We compared the impact of different prevention measures and their combinations on these outputs by varying the reduction in infectivity of disease-aware infectious individuals (mask-wearing), the reduction in susceptibility of disease-aware susceptible individuals (hand washing), the reduction in contact rate of disease-aware individuals only (self-imposed social distancing), and of all individuals (government-imposed social distancing).

We refer to these quantities as the efficacy of a prevention measure and vary it from 0% (zero efficacy) to 100% (full efficacy) (Table 1).

The main analyses were performed for two values of the rate of awareness spread that corresponded to scenarios of slow and fast spread of awareness in the population (Table 1). For these scenarios, the proportions of the aware population at the peak of the epidemic were 40% and 90%, respectively. In the main analyses, government-imposed social distancing was initiated when 10 individuals got diagnosed and was lifted after 3 months.

Table 1. Parameter values for the transmission model with and without awareness.

Parameters		Value*	Source
Epidemiological parameters			
Basic reproduction number	R_0	2.5 (2–3)	Li and colleagues [5], Park and colleagues [30], sensitivity analyses
Probability of transmission per contact with I_S	ε	0.048	From $R_0 = \beta[p\sigma/\gamma_M+(1-p)/\nu]$
Transmission rate of infection via contact with I_S	Β	0.66 per day	$\beta = c\varepsilon$
Average contact rate (unique persons)	C	13.85 persons per day	Mossong and colleagues [31]
Relative infectivity of infectious with mild disease (I_M)	σ	50% (25%–75%)	Assumed, see, e.g., Liu and colleagues [29], sensitivity analyses
Proportion of infectious with mild disease (I_M)	P	82% (82%–90%)	Wu and colleagues [32], Anderson and colleagues [20], sensitivity analyses
Delay between infection and onset of infectiousness (latent period)	$1/\alpha$	4 days	Shorter than incubation period [5, 30, 33]
Delay from onset of infectiousness to diagnosis for I_S	$1/\nu$	5 (3–7) days	Li and colleagues [5], sensitivity analyses
Recovery period of infectious with mild disease (I_M)	$1/\gamma_M$	7 (5–9) days	Li Xingwang†, sensitivity analyses
Delay from diagnosis to recovery for unaware diagnosed (I_D)	$1/\gamma_S$	14 days	WHO [34]
Relative infectivity of isolated (I_D)		0%	Assuming perfect isolation
Case fatality rate of unaware diagnosed (I_D)	f	1.6%	Althaus and colleagues [35] Park and colleagues [30]
Disease-associated death rate of unaware diagnosed (I_D)	η	0.0011 per day	$\eta = \gamma_S f/(1-f)$
Awareness parameters			

Rate of awareness spread (slow, fast and range)	δ	5×10^{-5}, 1 (10^{-6} – 1) per year	Assumed, sensitivity analyses
Relative susceptibility to awareness acquisition for S, E, I_M, and R_M	k	50% (0%–100%)	Assumed, sensitivity analyses
Duration of awareness for S^a, E^a, IaM, and RaM	$1/\mu$	30 (7–365) days	Assumed, sensitivity analyses
Duration of awareness for IaS	$1/\mu_S$	60 (7–365) days	Longer than $1/\mu$, sensitivity analyses
Delay from onset of infectiousness to diagnosis for IaS	$1/\nu^a$	3 (1–5) days	Shorter than $1/\nu$, sensitivity analyses
Delay from diagnosis to recovery of aware diagnosed (IaD)	1/γaS	12 days	Shorter than $1/\gamma_S$
Case fatality rate of aware diagnosed (IaD)	f^a	1%	Smaller than f
Disease-associated death rate of aware diagnosed (IaD)	η^a	0.0008 per day	η=γaSfa/(1−fa)
Prevention measure parameters			
Efficacy of mask-wearing (reduction in infectivity)		0%–100%	Varied
Efficacy of hand washing (reduction in susceptibility)		0%–100%	Varied
Efficacy of self-imposed contact rate reduction		0%–100%	Varied
Efficacy of government-imposed contact rate reduction		0%–100%	Varied
Duration of government-imposed social distancing		3 (1–13) months	Assumed, sensitivity analyses
Threshold for initiation of government-imposed social distancing		10 (1–1,000) diagnoses	Assumed, sensitivity analyses

*Mean or median values were used from literature; range was used in the sensitivity analyses.

†Expert at China's National Health Commission.

Estimates of epidemiological parameters were obtained from the most recent literature (Table 1). We used contact rates for the Netherlands, but the model is appropriate for other Western countries with similar contact patterns. A detailed mathematical description of the model can be found. The model was implemented in Mathematica 10.0.2.0. The code reproducing the results of this study is available at https://github.com/lynxgav/COVID19-mitigation.

Sensitivity analyses

To allow for the uncertainty in the parameters of the baseline transmission model, we conducted sensitivity analyses with respect to the proportion of infectious individuals with mild disease, the relative infectivity of infectious individuals with mild disease, the recovery period of infectious individuals with mild disease, the delay from onset of infectiousness to diagnosis for infectious individuals with severe disease, and the basic reproduction number.

We also conducted sensitivity analyses for the model with disease awareness with respect to changes in the delay from the onset of infectiousness to diagnosis and isolation for disease-aware individuals, the rate of awareness spread, the relative susceptibility to awareness, and the duration of awareness. Parameter ranges used in these sensitivity analyses are specified in Table 1.

In addition, we present results for the impact on the model outcomes of all combinations of self-imposed prevention measures as their efficacy was varied from 0% to 100% and of the government-imposed social distancing, with efficacy ranging from 0% to 100%, different thresholds for initiating the intervention (1 to 1,000 diagnoses), and different durations of the intervention (3, 8, and 13 months).

RESULTS

Our analyses show that disease awareness spread has a significant effect on the model predictions. We first considered the epidemic dynamics in a disease-aware population where hand washing is promoted, as an example of self-imposed measures (Fig 1).

Then, we performed a systematic comparison of the impact of different prevention measures on the model output for slow (Fig 2) and fast (Fig 3) rate of awareness spread.

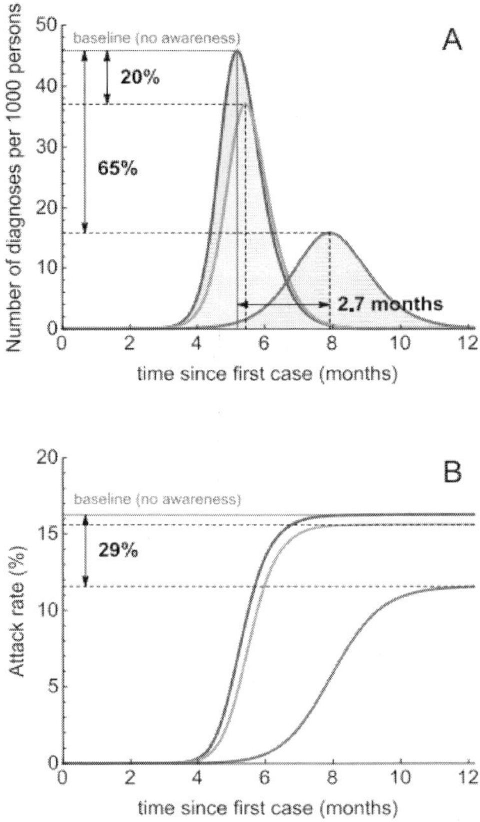

Figure 3. Illustrative simulations of the transmission model.

(A, B) Shows the number of diagnoses and the attack rate during the first 12 months after the first case under three model scenarios. The red lines correspond to the baseline transmission model. The orange lines correspond to the model with a fast rate of awareness spread and no interventions. The blue lines correspond to the latter model, where disease awareness induces the uptake of hand washing with an efficacy of 30%.

Epidemic dynamics

All self-imposed measures and government-imposed social distancing have an effect on the COVID-19 epidemic dynamics. The qualitative and quantitative impact, however, depends strongly on the prevention measure and the rate of awareness spread. The baseline model predicts 46 diagnoses per 1,000 individuals at the peak of the epidemic, an attack rate of about 16%, and the time to the peak of about 5.2 months (red line, Fig 3A and 3B). In the absence of prevention measures, a fast spread of disease awareness reduces the peak number of diagnoses by 20% but has only a minor effect on the attack rate and peak timing (orange line, Fig 3A and 3B). This is expected, as disease-aware individuals with severe disease seek medical care sooner and therefore get diagnosed faster, causing fewer new infections as compared to the baseline model. Awareness dynamics coupled with the use of self-imposed prevention measures has an even larger impact on the epidemic. The blue line in Fig 3A shows the epidemic curve for the scenario when disease-aware individuals use hand washing as a self-imposed prevention measure. Even if the efficacy of hand washing is modest (i.e., 30% as in Fig 3A), the impact on the epidemic can be significant; namely, we predicted a 65% reduction in the peak number of diagnoses, a 29% decrease in the attack rate, and a delay in peak timing of 2.7 months (Fig 3A and 3B).

The effect of awareness on the disease dynamics can also be observed in the probability of infection during the course of the epidemic. In the model with awareness and no measures, the probability of infection is reduced by 4% for all individuals. Hand washing with an efficacy of 30% reduces the respective probability by 14% for unaware individuals and by 29% for aware individuals. Note that the probability of infection is highly dependent on the type of prevention measure.

A comparison of prevention measures

Slow spread of awareness

Fig 4 shows the impact of all considered self-imposed measures as well as of the government-imposed social distancing on the peak number of diagnoses, attack rate, and the time to the peak for a slow rate of awareness spread. In this scenario, the model predicts progressively larger reductions in the peak number of diagnoses and in the attack rate as the efficacy of the self-imposed measures increases. In the limit of 100% efficacy, the reduction in the peak number of diagnoses is 23% to 30% (Fig 4A) and the attack rate decreases from 16% to 12%–13% (Fig 4B). The efficacy of the self-imposed measures has very little impact on the peak timing when compared to the baseline; i.e., no awareness in the population (Fig 4C). Because the proportion of aware individuals who change their behavior is too small to make a significant impact on transmission, self-imposed measures can only mitigate but not prevent an epidemic.

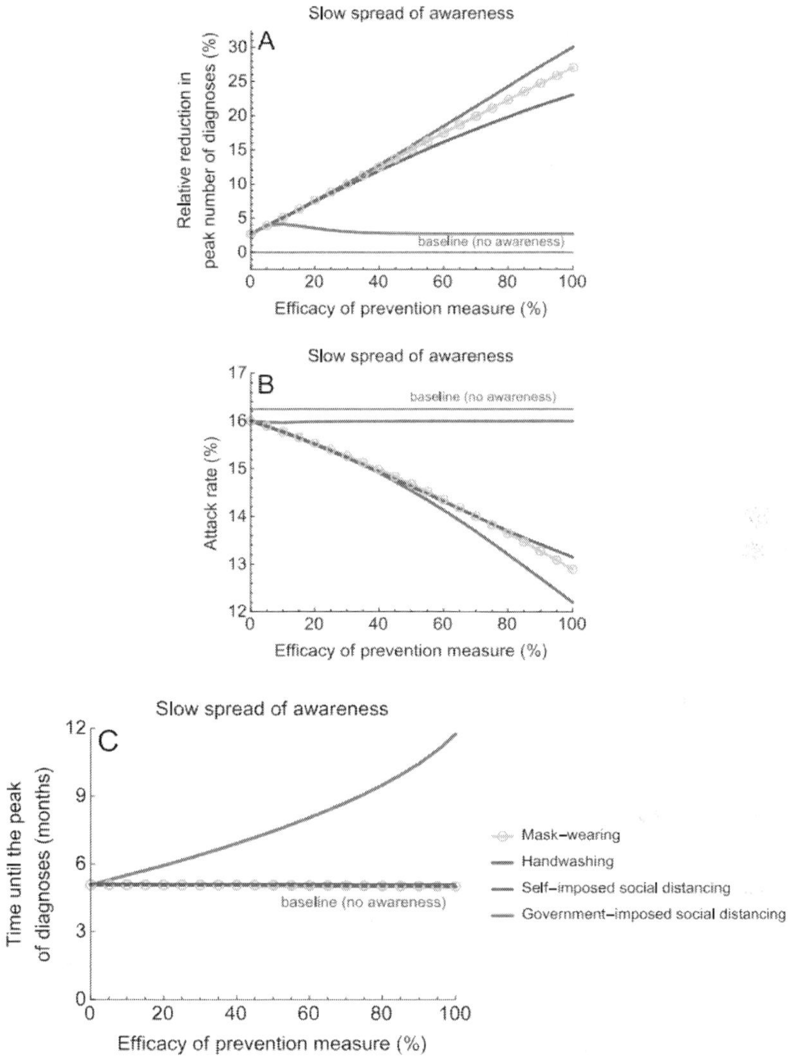

Figure 4. Impact of prevention measures on the epidemic for a slow rate of awareness spread.

(A–C) Shows the relative reduction in the peak number of diagnoses, the attack rate (proportion of the population that recovered or died after severe infection), and the time until the peak number of diagnoses. The efficacy of prevention measures was varied between 0% and 100%. In the context of this study, the efficacy of social distancing denotes the reduction in the contact rate. The efficacy of hand washing and mask-wearing are given by the reduction in susceptibility and infectivity, respectively. The simulations were started with one case. Government-imposed social distancing was initiated after 10 diagnoses and lifted after 3 months. For parameter values, see Table 1. Please note that the blue line corresponding to hand washing is not visible in (C) because it almost completely overlaps with lines for mask-wearing and self-imposed social distancing.

When awareness spreads at a slow rate, a 3-month government intervention has a contrasting impact to the self-imposed measure scenario. The time to the peak number of diagnoses is longer for more stringent contact rate reductions.

For example, a complete lockdown (government-imposed social distancing with 100% efficacy) can postpone the peak by almost 7 months, but its magnitude and attack rate are unaffected (with respect to the baseline model without measures and awareness).

Similar predictions are expected, as long as government-imposed social distancing starts early (e.g., after tens to hundreds of cases) and is lifted a few weeks to a few months later. This type of intervention halts the epidemic for the duration of intervention, but, because of a large pool of susceptible individuals, epidemic resurgence is expected as soon as social distancing measures are lifted.

Fast spread of awareness

Because the government intervention reduces the contact rate of all individuals irrespective of their awareness status, it has a comparable impact on transmission for scenarios with fast and slow rate of awareness spread (compare Fig 4 and Fig 5).

-However, the impact of self-imposed measures is drastically different when awareness spreads fast. All self-imposed measures are more effective than the short-term government intervention. These measures not only reduce the attack rate (Fig 5B) and diminish and postpone the peak number of diagnoses (Fig 5A and 5C), but they can also prevent a large epidemic altogether when their efficacy is sufficiently high (about 50%). Note that when the rate of awareness is fast, as the number of diagnoses grows, the population becomes almost homogeneous, with most individuals being disease-aware. It can be shown that in such populations, prevention measures yield comparable results if they have the same efficacy.

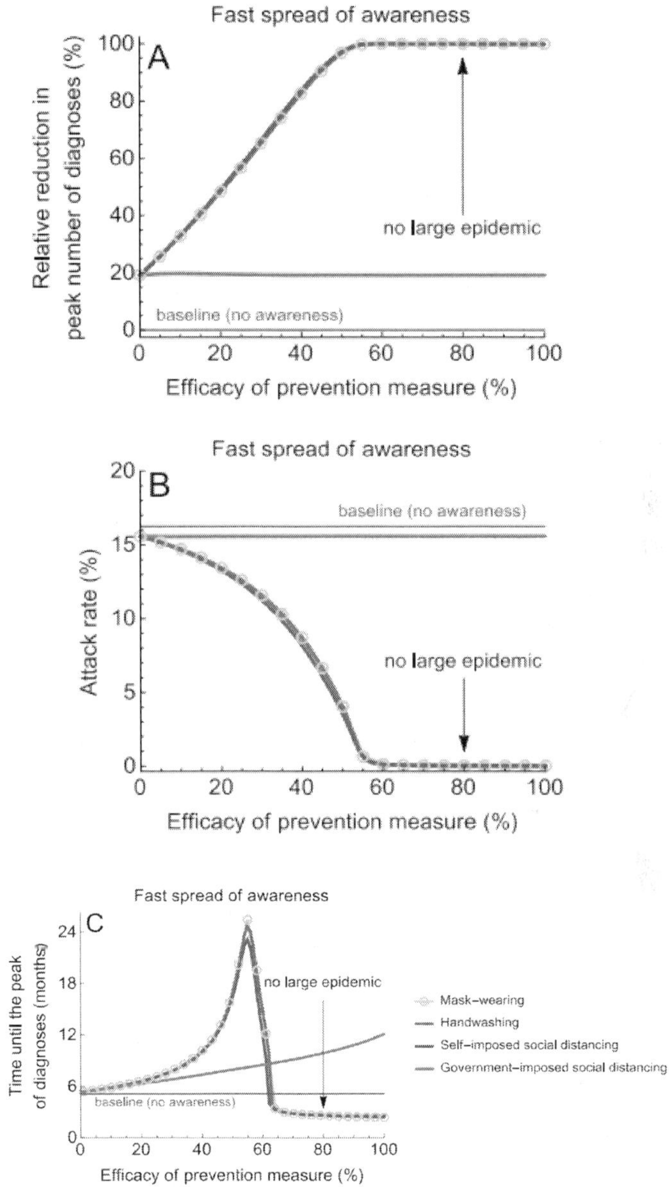

Figure 5. Impact of prevention measures on the epidemic for a fast rate of awareness spread.

(A–C) Shows the relative reduction in the peak number of diagnoses, the attack rate (proportion of the population that recovered or died after severe infection), and the time until the peak number of diagnoses. The efficacy of prevention measures was varied between 0% and 100%. In the context of this study, the efficacy of social distancing denotes the reduction in the contact rate. The efficacy of hand washing and mask-wearing are given by the reduction in susceptibility and infectivity, respectively. The simulations were started with one case. Government-imposed social distancing was initiated after 10 diagnoses and lifted after 3 months. For parameter values, see Table 1. Please note that the blue line corresponding to hand washing is not visible in (A) because it almost completely overlaps with lines for mask-wearing and self-imposed social distancing.

Combinations of prevention measures

If government-imposed social distancing is combined with a self-imposed prevention measure, the model predicts that the relative reductions in the peak number of diagnoses and attack rate are determined by the efficacy of the self-imposed measure, whereas the timing of the peak is determined by the efficacies of both the self-imposed measure and the government intervention.

This is demonstrated in Fig 6, where we used a combination of hand washing with efficacies of 30%, 45%, and 60%, and government-imposed social distancing with efficacy ranging from 0% to 100% for slow and fast spread of awareness.

Our results show that the effect of the combined intervention highly depends on the rate of awareness spread. Fast awareness spread is crucial for a large reduction in the peak number of diagnoses (Fig 6A) and in the attack rate (Fig 6B).

Note, that for fast spread of awareness, a combination of a complete lockdown and hand washing with an efficacy of 30% could postpone the time to the peak number of diagnoses by nearly 10 months (Fig 6C). Thus, when combined with short-term government-imposed social distancing, hand washing can contribute to mitigating and delaying the epidemic after the lockdown is relaxed.

The second wave of the epidemic could be prevented completely if the efficacy of hand washing exceeds 50% (Fig 6A). The results for the combination of mask-wearing and government-imposed social distancing are similar.

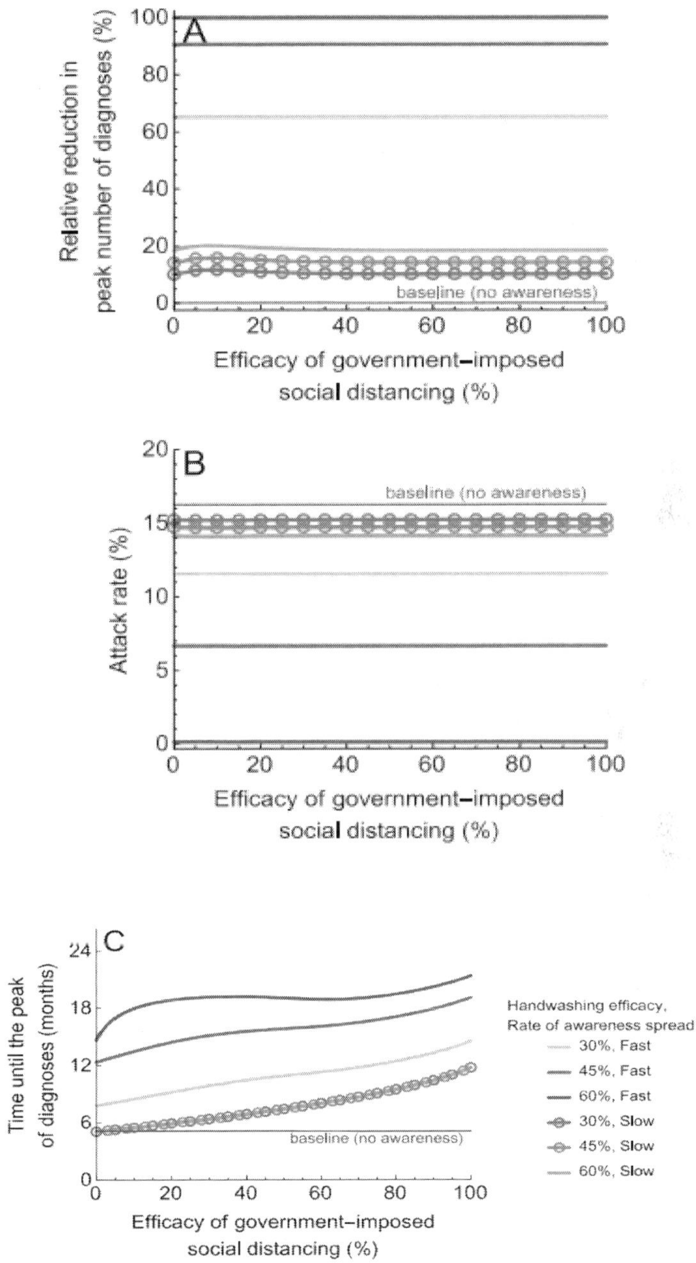

Figure 6. Impact on the epidemic of a combination of government-imposed social distancing and hand washing.

(A–C) Shows the relative reduction in the peak number of diagnoses, the attack rate (proportion of the population that recovered or died after severe infection), and the time until the peak number of diagnoses. The efficacy of hand washing was 30%, 45%, and 60%. In the context of this study, the efficacy of social distancing denotes the reduction in the contact rate. The efficacy of hand washing is given by the reduction in susceptibility. The simulations were started with one case. Government-imposed social distancing was initiated after 10 diagnoses and lifted after 3 months. For parameter values, see Table 1.

The effect of combinations of self-imposed measures (e.g., hand washing and mask-wearing) is additive. This means that, for fast spread of awareness, a large outbreak can be prevented by, for example, a combination of hand washing and self-imposed social distancing, each with an efficacy of around 25% (or other efficacies adding up to 50%).

DISCUSSION

For many countries around the world, the focus of public health officers in the context of COVID-19 epidemic has shifted from containment to mitigation and delay. Our study provides new insights for designing effective outbreak control strategies. Based on our results, we conclude that hand washing, mask-wearing, and social distancing adopted by disease-aware individuals can delay the epidemic peak, flatten the epidemic curve, and reduce the attack rate. We show that the rate at which disease awareness spreads has a strong impact on how self-imposed measures affect the epidemic. For a slow rate of awareness spread, self-imposed measures have less impact on transmission, as not many individuals adopt them. However, for a fast rate of awareness spread, their impact on the magnitude and timing of the peak increases with increasing efficacy of the respective measure. For all measures, a large epidemic can be prevented when the efficacy exceeds 50%. Moreover, the effect of combinations of self-imposed measures is additive. In practical terms, it means that SARS-CoV-2 will not cause a large outbreak in a country where 90% of the population adopts hand washing and social distancing that are 25% efficacious (i.e., reduce susceptibility and contact rate by 25%, respectively).

Although our analyses indicate that the effects of self-imposed measures on mitigating and delaying the epidemic for the same efficacies are similar (see Fig 4 and Fig 5), not all explored efficacy values may be achieved for each measure. Wong and colleagues [22] and Cowling and colleagues

[24] performed a systematic review and meta-analysis on the effect of hand washing and face masks on the risk of influenza virus infections in the community. While the authors highlight the potential importance of both hand hygiene and face masks, only modest effects could be ascertained with a pooled risk ratio of 0.73 (95% CI 0.6–0.89) for a combination of these two measures. However, the authors also highlight the small number of randomized controlled trials and the heterogeneity of the studies as notable limitations that may have led to these results. Given the high uncertainty around the efficacies of hand hygiene and mask-wearing on their own, the promotion of a combination of these measures might become preferable to recommending hand washing or mask-only measures. For self-imposed social distancing, contacts might not be eliminated entirely (e.g., household contacts remain), and therefore realistic values of the efficacy of self-imposed social distancing can be close to but may never reach 100%. Thus, for a fair comparison between measures, realistic efficacy values of a specific measure should be taken into consideration.

We contrasted self-imposed measures stimulated by disease awareness with mandated social distancing. Our analyses show that short-term government-imposed social distancing that is implemented early into the epidemic can delay the epidemic peak but does not affect its magnitude nor the attack rate. For example, a complete lockdown of 3 months imposing a community-wide contact rate reduction that starts after tens to thousands diagnoses in the country can postpone the peak by about 7 months. Such an intervention is highly desirable, when a vaccine is being developed or when healthcare systems require more time to treat cases or increase capacity. If this intervention is implemented in a population that exercises a self-imposed measure that continues to be practiced even after the lockdown is over, then the delay can be even longer (e.g., up to 10 months for hand washing with 30% efficacy). In the context of countries that implemented social distancing as a measure to "flatten the curve" of the ongoing epidemics, peaked in cases, and are now planning or have already started gradual lifting of social distancing, it means that governments and public health institutions should intensify the promotion of self-imposed measures to diminish and postpone the peak of the potential second epidemic wave. The potential second wave could be prevented altogether if the coverage of a self-imposed measure in the population and its efficacy are sufficiently high (e.g., 90% and 50%, respectively). Our sensitivity analyses showed that lower or higher efficacies can be required to prevent a large epidemic for countries with smaller or larger basic reproduction numbers.

Since for many countries the COVID-19 epidemic is still in its early stages, government-imposed social distancing was modelled as a short-term intervention initiated when the number of diagnosed individuals was relatively low. Our sensitivity analyses showed that government interventions introduced later into the epidemic (at 100–1,000 diagnoses) and imposed for a longer period of time (3–13 months) not only delay the peak of the epidemic but also reduce it for intermediate efficacy values. Previous studies suggested that the timing of mandated social distancing is crucial for its viability in controlling a large disease outbreak [13, 14, 16, 38]. As discussed by Hollingsworth and colleagues [16] and Anderson and colleagues [20], a late introduction of such interventions may have a significant impact on the epidemic peak and attack rate. However, the authors also showed that the optimal strategy is highly dependent on the desired outcome. A detailed analysis of government intervention with different timings and durations that also takes into account the economic and societal consequences, and the cost of SARS-CoV-2 transmission is a subject for future work.

To our knowledge, our study is the first to provide comparative analysis of a suite of self-imposed measures, government-imposed social distancing, and their combinations as strategies for mitigating and delaying a COVID-19 epidemic. Several studies (e.g., [39–42]) looked at the effect of different forms of social distancing, but they did not include self-imposed measures such as hand washing and mask-wearing. Some of these studies concluded that one-time social-distancing interventions will be insufficient to maintain COVID-19 prevalence within the critical care capacity [40, 42]. In our analyses, we explored the full efficacy range for all self-imposed prevention measures and different durations and thresholds for initiation of government intervention. Our results allow drawing conclusions on which combination of prevention measures can be most effective in diminishing and postponing the epidemic peak when realistic values for the measure's efficacy are taken into account. We showed that spreading disease awareness such that highly efficacious preventive measures are quickly adopted by individuals can be crucial in reducing SARS-CoV-2 transmission and preventing a large epidemics of COVID-19.

Our model has several limitations. It does not account for stochasticity, demographics, heterogeneities in contact patterns, spatial effects, inhomogeneous mixing, imperfect isolation of individuals with severe disease, and reinfection with COVID-19. Our conclusions can therefore be drawn on a qualitative level. Detailed models will have to be developed to design and tailor effective strategies in particular settings. The impact of the duration of immunity has been explored by Kissler and colleagues [43]. The effect of nonpermanent immunity on the results of our model would be an interesting subject for future work. To take into account the uncertainty in SARS-CoV-19 epidemiological parameters, we performed sensitivity analyses to test the robustness of the model predictions. As more data become available, our model can be easily updated. In addition, our study assumes that individuals become disease-aware with a rate of awareness acquisition proportional to the number of currently diagnosed individuals. Other forms for the awareness acquisition rate that incorporate, e.g., the saturation of awareness, may be more realistic and would be interesting to explore in future studies. Furthermore, we assume that hand washing may reduce the susceptibility of an individual down to 0% and therefore neglect aerosol transmission of SARS-CoV-2. Thus, the impact of hand washing on the epidemic may be an overestimation. However, while there is preliminary evidence on SARS-CoV-2 RNA detection in aerosols [44], there is still uncertainty about the level of infectiousness of the detected aerosols and the significance of potential airborne transmission. Current recommendations by the World Health Organization are still focused on droplet and contact precautions [45]. Our model may be adapted as more information on the relative contribution of the transmission routes of COVID-19 emerges.

In conclusion, we provide the first empirical basis of how stimulating the uptake of effective prevention measures, such as hand washing or mask-wearing, combined with government-imposed social distancing intervention, can be pivotal to achieving control over a COVID-19 epidemic. While information on the rising number of COVID-19 diagnoses reported by the media may fuel anxiety in the population, wide and intensive promotion of self-imposed measures with proven efficacy by governments or public health institutions may be a key ingredient to tackle COVID-19.

ACKNOWLEDGMENTS

This study has benefited greatly from the feedback provided by the Infectious Disease Modelling group based at the Julius Center for Health Sciences and Primary Care, UMC Utrecht, Utrecht University.

FUNDING STATEMENT

This study was funded by the following: Fundação para a Ciência e a Tecnologia, project reference 131_596787873, awarded to GR, https://www.

fct.pt; ZonMw 91216062, awarded to MEK, funded MEK and AT, https://www.zonmw.nl/en/; One Health European Joint Programme Horizon 2020 project 773830 (award recipient is not an author of this manuscript) funded NGG and MCJB, https://ec.europa.eu/programmes/horizon2020/en; and Aidsfonds Netherlands project P-29704 (award recipient is not an author of this manuscript) funded GR, https://aidsfonds.nl/. The funders had no role in study design, data collection and analysis, decision to publish, or preparation of the manuscript.

REFERENCES

1. World Health Organization. Coronavirus disease 2019 (COVID-19). Situation Report–51; 2020 March 11. [cited 2020 Mar 13]. Available from: https://www.who.int/docs/default-source/coronaviruse/situation-reports/20200311-sitrep-51-covid-19.pdf.
2. Gostic KM, Kucharski AJ, Lloyd-Smith JO. Effectiveness of traveller screening for emerging pathogens is shaped by epidemiology and natural history of infection. eLife. 2015; 4:e05564.
3. Li R, Pei S, Chen B, Song Y, Zhang T, Yang W, et al. Substantial undocumented infection facilitates the rapid dissemination of novel coronavirus (SARS-CoV2). Science. 2020; 368(6490):489–493. 10.1126/science.abb3221
4. Tindale L, Coombe M, Stockdale JE, Garlock E, Lau WYV, Saraswat M, et al. Transmission interval estimates suggest pre-symptomatic spread of COVID-19. medRxiv:2020.03.03.20029983 [Preprint]. 2020. [cited 2020 Jun 25]. https://www.medrxiv.org/content/10.1101/2020.03.03.20029983v1
5. Li Q, Guan X, Wu P, Wang X, Zhou L, Tong Y, et al. Early transmission dynamics in Wuhan, China, of novel coronavirus-infected pneumonia. N Engl J Med 2020; 382:1199–1207. 10.1056/NEJMoa2001316
6. European Centre for Disease Prevention and Control. Coronavirus disease 2019 (COVID-19) pandemic: increased transmission in the EU/EEA and the UK–seventh update; 2020 March 25. [cited 2020 Apr 26]. Available from: https://www.ecdc.europa.eu/sites/default/files/documents/RRA-seventh-update-Outbreak-of-coronavirus-disease-COVID-19.pdf
7. European Centre for Disease Prevention and Control. Outbreak of acute respiratory syndrome associated with a novel coronavirus, China: first local transmission in the EU/EEA—third update. 2020 January 31. [cited 2020 Jun 25]. Available from: https://www.ecdc.europa.eu/sites/default/files/documents/novel-coronavirus-risk-assessment-china-31-january-2020_0.pdf.
8. Ferretti L, Wymant C, Kendall M, Zhao L, Nurtay A, Abeler-Dörner L, et al. Quantifying SARS-CoV-2 transmission suggests epidemic control with digital contact tracing. Science. 2020; 368(6491):eabb6936 10.1126/science.abb6936
9. Hellewell J, Abbott S, Gimma A, Bosse NI, Jarvis CI, Russell TW, et

al. Feasibility of controlling COVID-19 outbreaks by isolation of cases and contacts. Lancet Glob Health. 2020;8(4):e488–e496. 10.1016/S2214-109X(20)30074-7

10. Fraser C, Riley S, Anderson RM, Ferguson NM. Factors that make an infectious disease outbreak controllable. Proc Natl Acad Sci U S A. 2004;101(16):6146–6151. 10.1073/pnas.0307506101
11. Klinkenberg D, Fraser C, Heesterbeek H. The effectiveness of contact tracing in emerging epidemics. PLoS ONE. 2006;1(1):e12.
12. Wilder-Smith A, Freedman DO. Isolation, quarantine, social distancing and community containment: pivotal role for old-style public health measures in the novel coronavirus (2019-nCoV) outbreak. J Travel Med. 2020. 10.1093/jtm/taaa020
13. Hatchett RJ, Mecher CE, Lipsitch M. Public health interventions and epidemic intensity during the 1918 influenza pandemic. Proc Natl Acad Sci U S A. 2007;104(18):7582–7587. 10.1073/pnas.0610941104
14. Bootsma MCJ, Ferguson NM. The effect of public health measures on the 1918 influenza pandemic in U.S. cities. Proc Natl Acad Sci U S A. 2007;104(18):7588–7593. 10.1073/pnas.0611071104
15. Markel H, Lipman HB, Navarro JA, Sloan A, Michalsen JR, Stern AM, et al. Nonpharmaceutical interventions implemented by US cities during the 1918–1919 influenza pandemic. JAMA. 2007;298(6):644–654. 10.1001/jama.298.6.644
16. Hollingsworth TD, Klinkenberg D, Heesterbeek H, Anderson RM. Mitigation strategies for pandemic influenza A: balancing conflicting policy objectives. PLoS Comput Biol. 2011;7(2):1–11.
17. Cauchemez S, Ferguson NM, Wachtel C, Tegnell A, Saour G, Duncan B, et al. Closure of schools during an influenza pandemic. Lancet Infect Dis. 2009;9:473–481. 10.1016/S1473-3099(09)70176-8
18. Jackson C, Vynnycky E, Hawker J, Olowokure B, Mangtani P. School closures and influenza: systematic review of epidemiological studies. BMJ Open. 2013;3(2).
19. European Centre for Disease Prevention and Control. Guidelines for the use of non-pharmaceutical measures to delay and mitigate the impact of 2019-nCoV. 2020 Feb 10. [cited 2020 Mar 18]. Available from: https://www.ecdc.europa.eu/en/publications-data/guidelines-use-non-pharmaceutical-measures-delay-and-mitigate-impact-2019-ncov.

20. Anderson RM, Heesterbeek H, Klinkenberg D, Hollingsworth TD. How will country-based mitigation measures influence the course of the COVID-19 epidemic? Lancet. 2020; 395(10228):P931–934.
21. Kampf G. Efficacy of ethanol against viruses in hand disinfection. J Hosp Infect. 2018;98(4):331–338. 10.1016/j.jhin.2017.08.025
22. Wong VWY, Cowling BJ, Aiello AE. Hand hygiene and risk of influenza virus infections in the community: a systematic review and meta-analysis. Epidemiology and infection. 2014;142(5):922–932. 10.1017/S095026881400003X
23. Institute of Medicine (IOM). Preventing transmission of pandemic influenza and other viral respiratory diseases: Personal protective equipment for healthcare personnel. Update 2010. Washington, DC: The National Academies Press; 2011.
24. Cowling BJ, Zhou Y, Ip DKM, Leung GM, Aiello AE. Face masks to prevent transmission of influenza virus: a systematic review. Epidemiology and Infection. 2010;138(4):449–456. 10.1017/S0950268809991658
25. Leung NHL, Chu DKW, Shiu EYC, Chan KH, McDevitt JJ, Hau BJP, et al. Respiratory virus shedding in exhaled breath and efficacy of face masks. Nat Med. 2020;26: 676–680. 10.1038/s41591-020-0843-2
26. Funk S, Gilad E, Watkins C, Jansen VAA. The spread of awareness and its impact on epidemic outbreaks. Proc Natl Acad Sci U S A. 2009;106(16):6872–6877. 10.1073/pnas.0810762106
27. Perra N, Balcan D, Gonçalves B, Vespignani A. Towards a characterization of behavior-disease models. PLoS ONE. 2011;6(8):1–15.
28. He D, Dushoff J, Day T, Ma J, Earn DJD. Inferring the causes of the three waves of the 1918 influenza pandemic in England and Wales. Proc. R. Soc. B. Biol Sci. 2013;280(1766):20131345.
29. Liu Y, Yan LM, Wan L, Xiang TX, Le A, Liu JM, et al. Viral dynamics in mild and severe cases of COVID-19. Lancet Infect Dis. 2020. 10.1016/S1473-3099(20)30232-2
30. Park M, Cook AR, Lim JT, Sun Y, Dickens BL. A systematic review of COVID-19 epidemiology based on current evidence. J Clin Med. 2020;9(4):967.
31. Mossong J, Hens N, Jit M, Beutels P, Auranen K, Mikolajczyk R, et al. Social contacts and mixing patterns relevant to the spread of infectious

diseases. PLoS Med. 2008;5(3):1–1.

32. Wu Z, McGoogan JM. Characteristics of and important lessons from the coronavirus disease 2019 (COVID-19) outbreak in China: summary of a report of 72 314 cases from the Chinese Center for Disease Control and Prevention. JAMA. 2020;323(13):1239–1242.
33. Backer JA, Klinkenberg D, Wallinga J. Incubation period of 2019 novel coronavirus (2019-nCoV) infections among travellers from Wuhan, China, 20–28 January 2020. Euro Surveill. 2020;25(5):pii = 2000062.
34. World Health Organization. WHO Director-General's opening remarks at the media briefing on COVID-19–24 February 2020; 2020 February 23. [cited 2020 Jun 25]. Available from: https://www.who.int/dg/speeches/detail/who-director-general-s-opening-remarks-at-the-media-briefing-on-covid-19—24-february-2020
35. Riou J, Counotte M, Hauser A, Althaus C. Adjusted age-specific case fatality ratio during the COVID-19 epidemic in Hubei, China, January and February 2020. medRxiv 2020.03.04.20031104 [Preprint]. 2020. [cited 2020 Mar 13]. Available from: https://www.medrxiv.org/content/10.1101/2020.03.04.20031104v2
36. World Health Organization. Coronavirus disease (COVID-19) advice for the public: When and how to use masks. 2020. [cited 2020 Mar 13]. Available from: https://www.who.int/emergencies/diseases/novel-coronavirus-2019/advice-for-public/when-and-how-to-use-masks.
37. Jarvis CI, Van Zandvoort K, Gimma A, Prem Ka, Klepac P, Rubin GJ, et al. Quantifying the impact of physical distance measures on the transmission of COVID-19 in the UK. BMC Med 2020;18(124). 10.1186/s12916-020-01597-8
38. Lauro, Francesco Di and Kiss, István Z and Miller, Joel. The timing of one-shot interventions for epidemic control. medRxiv 2020.03.02.20030007 [Preprint]. 2020. [cited 2020 May 5]. Available from: https://www.medrxiv.org/content/10.1101/2020.03.02.20030007v1.
39. Prem K, Liu Y, Russell TW, Kucharski AJ, Eggo RM, Davies N, et al. The effect of control strategies to reduce social mixing on outcomes of the COVID-19 epidemic in Wuhan, China: a modelling study. Lancet Public Health. 2020;5(5):e261 –e270. 10.1016/S2468-2667(20)30073-6
40. Report 9—Impact of non-pharmaceutical interventions (NPIs) to

reduce COVID-19 mortality and healthcare demand. 2020 [cited 2020 May 5]. Available from: https://www.imperial.ac.uk/media/imperial-college/medicine/sph/ide/gida-fellowships/Imperial-College-COVID19-NPI-modelling-16-03-2020.pdf.

41. Davies NG and Kucharski AJ and Eggo RMand Gimma Aand CMMID COVID-19 Working Group, Edmunds WJ. The effect of non-pharmaceutical interventions on COVID-19 cases, deaths and demand for hospital services in the UK: a modelling study. medRxiv 2020.04.01.20049908 [Preprint]. 2020. [cited 2020 May 5]. Available from: https://www.medrxiv.org/content/10.1101/2020.04.01.20049908v1.

42. Kissler SM and Tedijanto C and Lipsitch Mand Grad Y. Social distancing strategies for curbing the COVID-19 epidemic. medRxiv 2020.03.22.20041079 [Preprint]. 2020. [cited 2020 May 5]. Available from: https://www.medrxiv.org/content/10.1101/2020.03.22.20041079v1.

43. Kissler SM, Tedijanto C, Goldstein E, Grad YH, Lipsitch M. Projecting the transmission dynamics of SARS-CoV-2 through the postpandemic period. Science. 2020. 10.1126/science.abb5793

44. Liu Y, Ning Z, Chen Y, Guo M, Liu Y, Gali NK, et al. Aerodynamic analysis of SARS-CoV-2 in two Wuhan hospitals. Nature. 2020. 10.1038/s41586-020-2271-3

45. Modes of transmission of virus causing COVID-19: implications for IPC precaution recommendations. 2020 [cited 2020 May 5]. Available from: https://www.who.int/news-room/commentaries/detail/modes-of-transmission-of-virus-causing-covid-19-implications-for-ipc-precaution-recommendations.

11

Middle East Respiratory Syndrome Coronavirus (MERS-CoV): Infection, Immunological Response, and Vaccine Development

Ayman Mubarak,[1] Wael Alturaiki,[2] and Maged Gomaa Hemida[3,4]

[1]Department of Botany and Microbiology, College of Science, King Saud University, Saudi Arabia

[2]Department of Medical Laboratory Sciences, College of Applied Medical Sciences, Majmaah University, Majmaah 11952, Saudi Arabia

[3]Department of Microbiology and Parasitology, College of Veterinary Medicine, King Faisal University, Saudi Arabia

[4]Department of Virology, Faculty of Veterinary Medicine, Kafresheikh University, Egypt

ABSTRACT

Middle East respiratory syndrome coronavirus (MERS-CoV) first emerged in late 2012. Since its emergence, a total of 2279 patients from 27 countries have been infected across the globe according to a World Health Organization

Citation: Middle East Respiratory Syndrome Coronavirus (MERS-CoV): Infection, Immunological Response, and Vaccine Development. Ayman Mubarak, Wael Alturaiki, and Maged Gomaa Hemida,. Hindawi Journal of Immunology Research Volume 2019, Article ID 6491738, 11 pages https://doi.org/10.1155/2019/6491738

(WHO) report (Feb. 12th, 2019). Approximately 806 patients have died. The virus uses its spike proteins as adhesive factors that are proinflammatory for host entry through a specific receptor called dipeptidyl peptidase-4 (DPP4). This receptor is considered a key factor in the signaling and activation of the acquired and innate immune responses in infected patients. Using potent antigens in combination with strong adjuvants may effectively trigger the activation of specific MERS-CoV cellular responses as well as the production of neutralizing antibodies. Unfortunately, to date, there is no effective approved treatment or vaccine for MERS-CoV. Thus, there are urgent needs for the development of novel MERS-CoV therapies as well as vaccines to help minimize the spread of the virus from infected patients, thereby mitigating the risk of any potential pandemics. Our main goals are to highlight and describe the current knowledge of both the innate and adaptive immune responses to MERS-CoV and the current state of MERS-CoV vaccine development. We believe this study will increase our understanding of the mechanisms that enhance the MERS-CoV immune response and subsequently contribute to the control of MERS-CoV infections.

INTRODUCTION

Middle East respiratory syndrome coronavirus (MERS-CoV) is a novel human coronavirus that was previously called "novel human coronavirus Erasmus Medical Center" (HCoV-EMC). The virus was discovered for the first time in Saudi Arabia in 2012 by Zaki et al. [1]. The World Health Organization (WHO) has confirmed 2279 cases of human infections with MERS-CoV in 27 countries since 2012; 806 (35%) infected patients have died as of Feb. 13, 2019. However, Saudi Arabia still has the highest reported MERS-CoV mortality rate. Approximately 80% of the cases have been reported to occur there [2]. MERS-CoV belongs to the family *Coronaviridae*, order *Nidovirales*. It is one of the recently reported zoonotic viruses. The family *Coronaviridae* is classified into four genera (α, β, γ, and δ). Each genus is divided into linage subgroups. MERS-CoV belongs to lineage-C of the β coronaviruses [3, 4]. Although bats are the main reservoir for most coronaviruses, dromedary camels are considered the only known reservoir for MERS-CoV to date. Additionally, MERS-CoV isolated from dromedary camels is relatively closely related to some bat coronaviruses [5–7]. According to the WHO, MERS-CoV transmission between humans is possible and occurs in Middle East countries and the Republic of Korea [2]. Viral spread has been observed among health-care workers and among individuals visiting MERS-CoV-positive patients. The control of some of these outbreaks has been achieved by the local center of disease control and prevention (CDC) [2]. Immunocompromised individuals as well as patients with comorbidities are the groups most prone to severe MERS-CoV infection, which may lead to death of these infected patients in many cases [8–10].

Three MERS-CoV proteins are expressed on the envelope of the virus: the surface spike protein (S), the membrane glycoprotein (M), and the envelope protein (E). The S protein is responsible for viral entry via attachment to and fusion with the host cell membrane. MERS-CoV host cell receptors were identified to be cluster of differentiation 26, also known as dipeptidyl peptidase-4 [11, 12]. The interaction of MERS-CoV S proteins with the DPP4 receptor not only facilitates viral access into the host cell but also triggers signals that induce the immunosuppression of infected patients, enabling viral replication and spread [13].

Despite ongoing research on the development of specific therapies or vaccines against MERS-CoV, there is currently no effective prophylaxis or therapy for MERS-CoV, which hinders the treatment or control of the viral infection. Understanding the mechanism of the immune response against MERS-CoV infection will make the development of effective vaccine candidates achievable, especially if the vaccine candidates are strong enhancers for both cellular and humoral immunity. In this review, we will discuss how innate immunity and acquired immunity respond to MERS-CoV infections in light of the most up-to-date literature in this field of research. Moreover, we highlight the most recent advances in the field of MERS-CoV vaccines

MERS-COV INNATE IMMUNITY

Dendritic cells (DCs) are important contributors to innate immunity and can trigger the production of large quantities of cytokines and chemokines. These cells have the ability to migrate from peripheral tissues to the lymphoid tissue to activate the T cell population [14]. Thus, DCs are considered potential targets for pathogen invasion, as they form bridges between innate and adaptive immunity [14]. Subsequently, both the T cell (cell-mediated immunity) and the B cell (humoral immunity) arms of the adaptive immune response are stimulated for a specific response [14]. The mechanisms of the immune response triggered by MERS-CoV infection and immune evasion strategies have not yet been fully studied. Interestingly, MERS-CoV has evolved strategies to manipulate innate immunity and prevent or block IFN production pathways [15]. This ability may contribute substantially to the high case-fatality rates of MERS-CoV-infected patients, especially those who are immunocompromised [9].

Pattern recognition receptors (PRRs), such as Toll-like receptors (TLRs) and retinoic acid inducible gene-I- (RIG-I-) like receptors, are essential mediators of the innate immune response [16–18]. After viral recognition by the TLR, one of the two different adaptor molecules is recruited—either MyD88 (myeloid differentiation primary response 88) or Toll/interleukin-1 receptor- (TIR-) domain-containing adapter-inducing interferon-β (TRIF). These molecules further activate the MAPK and NF-κB pathways that are responsible for promoting the production of proinflammatory cytokines and IFNs [19–21]. Meanwhile, the spike protein of MERS-CoV triggers the expression of some negative regulators of the TLR signaling pathways. This induction subsequently results in the expression of both IL-1R-associated kinase (IRAK-M) and peroxisome proliferator-activated receptor-γ (PPAR), which are negative regulators of IRF7, the transcription factor that induces the expression of IFN-α and IFN-β [13]. The long-term persistence of these negative regulators impairs the clearance of MERS-CoV infections; therefore, MERS-CoV persistence at the site of infection is established [22].

MyD88-dependent and TRIF-dependent signaling both use the TLR-4 pathway to activate downstream effectors [22]. However, mice lacking TLR-4 have more severe SARS-CoV infections than wild-type mice have. Thus, the protective signaling role through TLR-4/TRIF might be considered a distinctive feature in the pathogenesis of some coronaviruses [23]. We believe that using some special adjuvants as agonists for TLR-4 and TLR-3 plus the expressed MERS-CoV-S protein may help to improve the immunogenicity against MERS-CoV infection.

TLR-3 initiates the activation of interferon regulatory factors (i.e., IRF7 and IRF3) after binding with its ligand in a manner independent of MyD88 [20]. The TLR-3 agonist (poly IC) has recently been reported as a potential therapy for MER-CoV infection in a mouse model [24]. After the administration of poly IC, type 1 interferon expression is induced (IFN-β and IFN-α) [25] (Figure 1). Thus, different effectors, such as natural killer cells, CD8 T cells, and macrophages, are activated, and their antiviral effects are triggered [26, 27]. It is known that the proinflammatory cytokine response (such as TNF-α and IL6 production) against any infection has several drawbacks in the host, such as pathological damage to tissues

[28]. In contrast, the proinflammatory cytokine response can control viral dissemination. Understanding the TLR signaling pathways in the context of MERS-CoV infection will contribute to control of the viral infection, thereby mitigating the risk of its spread.

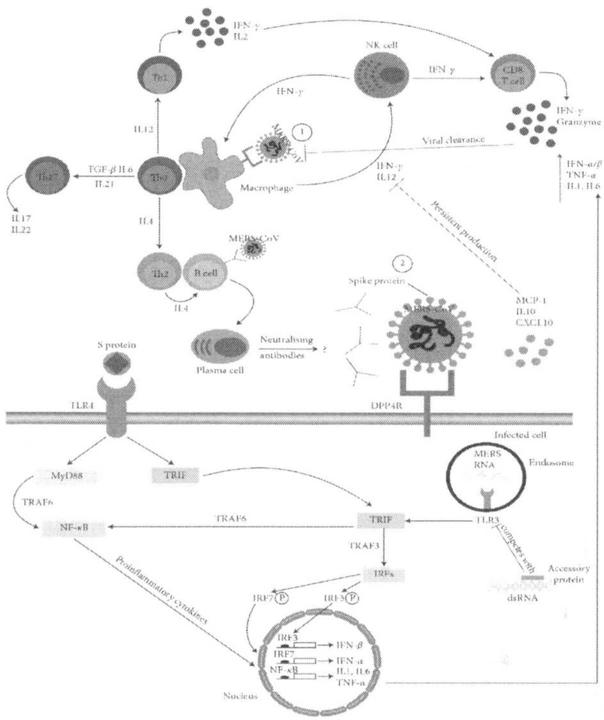

Figure 1. The proposed schematic representation of the immune response to MERS-CoV infection and how the invading virus is processed during an infection. (1) MERS-CoV infects macrophages through DPP4 binding, and then macrophages present MERS-CoV antigens to Th0 cells. This process leads to T cell activation and differentiation, including the production of cytokines associated with the different T cell subsets (i.e., Th1, Th2, and Th17), followed by a massive release of cytokines for immune response amplification. The continued production of these mediators due to viral persistence has a negative effect on Th0, NK, and CD8 T cell activation by inhibiting IL12 and IFN-γ production. However, CD8 T cells produce very effective mediators, such as IFN-γ and granzyme, to clear MERS-CoV. It is still unclear whether long-term or short-term protective antibodies are produced during neutralizing antibody production against MERS-CoV. (2) Attachment of MERS-CoV to DPP4 on the host cell through S protein leads to the appearance of genomic RNA in the cytoplasm.

An immune response to dsRNA can be partially generated during MERS-CoV replication. TLR-3 sensitized by dsRNA and cascades of signaling pathways (IRFs and NF-κB activation via TRAF3 and TRAF6, respectively) are activated to produce type I IFNs and proinflammatory cytokines. The production of type I IFNs is important to enhance the release of antiviral proteins for the protection of uninfected cells. Sometimes, accessory proteins of MERS-CoV can interfere with TLR-3 signaling and bind the dsRNA of MERS-CoV during replication to prevent TLR-3 activation and evade the immune response. TLR-4 might recognize S protein and lead to the activation of proinflammatory cytokines through the MyD88-dependent signaling pathway. Virus-cell interactions lead to strong production of immune mediators. The secretion of large quantities of chemokines and cytokines (MCP-1, IL10, and CXCL10) is promoted in infected cells in response to MERS-CoV infection. These chemokines and cytokines in turn recruit lymphocytes and leukocytes to the site of infection. Red arrows refer to inhibitory effects. Black arrows refer to activating effects.

Generally, IFNs play important roles during some viral infections and can be stimulated by double-stranded RNAs (dsRNAs) [15]. A study conducted by Chu et al. in 2014 demonstrated that monocyte-derived dendritic cells (Mo-DCs) infected with MERS-CoV exhibit no expression of IFN-β, despite the marginally early expression of IFN-α [29]. However, another recent study failed to stimulate the proinflammatory innate response and failed to produce type I IFNs in vitro in cultured infected cells, primary human airway epithelial cells and Mo-DCs infected with MERS-CoV [30, 31]. The mechanisms behind this response may be initially related to interference with the NF-κB signaling pathway, which is usually responsible for the induction of the proinflammatory response [32]. In addition, it is possible that the number of regulatory T cells induced at the site of infection has negative impacts on the expression of proinflammatory cytokines. Recently, treatment with IFN-α showed some promising trends in MERS-CoV-infected cells. The effect of the application of IFN-α in MERS-CoV-infected cells was 50-100-fold greater than that in SARS-CoV-infected cells [33]. Additionally, Hart et al. studied different IFN products and two different antiviral drugs, namely, ribavirin and mycophenolic acid (MPA), against MERS-CoV infection (Hu/Jordan-N3/2012) in vitro. The researchers showed that IFN-β has a potent inhibitory effect on MERS-CoV in comparison to that of other tested IFNs. Compared with ribavirin treatment, MPA treatment caused a stronger inhibition of viral replication in vitro, with an IC_{50} of 2.87 μM [34]. MPA was shown to enhance IFN-stimulated gene expression, suggesting that MPA is dependent on the modulation of the expression of IFN-stimulated genes [35]. Thus, MPA might provide an alternative treatment for MERS-CoV infection. IFN-β and MPA as combination or single therapies might provide great benefit as potent inhibitors in the treatment of MERS-CoV-infected patients by reducing viral loads. The FDA has approved the use of both IFN-β and MPA for other indications, and these therapies are currently in use [34].

MERS-CoV infects and replicates inside macrophages and subsequently induces the expression of MHC-I, MHC-II, and costimulation-related genes [28]. Some researchers have investigated the great impact of the MERS-CoV spike glycoprotein on the responsiveness of macrophages and monocytes (THP-1 cells) via TLR-4 signaling pathways [13]. They have shown that the MERS-CoV-S protein has a negative impact on the production of proinflammatory cytokines (IL6 and TNF-α). In contrast, this virus increases the production of anti-inflammatory cytokines, such as IL10. As suggested by Nicholls et al., the cytokines released by alveolar macrophages could have substantial effects on the pathogenicity of SARS-CoV [36].

MERS-CoV infection stimulates the production of type I IFNs (IFN-α and IFN-β) by infected cells, which leads to the release of some chemokines, such as MCP-1, CXCL10, and the cytokine IL10, which are responsible for T cell recruitment [37, 38]. It is known that CD4 helper T cells (T_h1) and natural killer (NK) cells require signaling by IL12 and IFN-γ for their activation. IFN-γ contributes to the activation of the two main arms of the immune response that help clear viruses: NK cells and $CD8^+$ T cells [26].

However, the persistence of MCP-1 and CXCL10 has a negative impact on the expression of IL12 and IFN [37] (Figure 1). Mahallawi et al. found no remarkable differences in the IL12 and Th2 cytokine expression profiles between MERS-CoV-infected patients and a healthy control group [39]. In the context of MERS-CoV infection, the production of both CXCL10 and IL10 increased in patients' sera within 0-3 days postinfection. However, the patients who did not recover or tolerate the infection had persistent viral replication due to the increase in the expression levels of both CXCL10 and IL10 [37]. Thus, these cytokines have a negative impact on the antiviral Th1-mediated effects [40]. Another study reported the upregulation of IL10 expression in MERS-CoV-infected patients compared to that in healthy volunteers [39]. IL10 has a positive impact on the production of proinflammatory cytokines mediated by the Janus kinase/signal transducer and activator of transcription (JAK-STAT) pathway [41]. This cytokine production is positively correlated with the Th2-mediated response (i.e., IL4 and IL13 expression), which in turn inhibits the type II IFN (IFN-γ) expression level [42]. The expression of this cytokine is also associated with persistence in some other viral infections, such as human immunodeficiency virus (HIV), hepatitis C virus (HCV), and hepatitis B virus (HBV) [43].

MERS-COV-ADAPTIVE IMMUNITY

MERS-CoV Cell-Mediated Response

T cells are the key players required for immunity against viral infections; $CD4^+$ T cells facilitate virus-specific antibody production through the T-dependent activation of B cells. However, $CD8^+$ T cells are cytotoxic and kill virus-infected cells [44]. Through the comparison of T cell-deficient BALB/c mice (mediated by the transduction of Ad5-hDPP4) with control mice and B cell-deficient mice, researchers determined that T cells are able to survive and destroy virus-infected cells in the infected lung [45]. This report may highlight the important roles of T cells but not B cells in controlling and fine-tuning the pathogenesis and outcomes of MERS-CoV infection. Zhao et al. infected Ad5-hDPP4-transduced BALB/c mice with either SARS-CoV or MERS-CoV. Subsequently, these mice were challenged with both viruses 5 weeks later. The results confirmed that the initial infection with SARS-CoV led to a significant decrease in MERS-CoV titers at day 5 post infection. Thus, a cross-reactive T cell response may result in decreasing MERS-CoV titers [45]. The roles of T and B cell responses in the context of MERS-CoV infection were studied. Both activated CD8 cells and anti-MERS-CoV antibodies were crucial for the clearance of the initial infection and protection against a subsequent challenge with the virus, respectively. This finding implies that the response to MERS-CoV generally occurs through antibody-mediated immunity [45]. Another study demonstrated that mice vaccinated with DNA encoding the modified SARS-CoV-S glycoprotein developed protective immunity resulting from the induction of T cells and the production of neutralizing antibodies. The protection was mainly due to an antibody-dependent (and not T cell-dependent) response [46]. Yang et al. reported that specific memory cells against spike proteins have no effect on viral clearance, even 2 days post challenge [46]. This result was confirmed when virus-specific T cells were depleted. However, this effect of cell depletion was not timely monitored at different intervals [47]. Hencc, the antiviral effects of the deplcted cells may be important during later infection time points, leading to the persistence of the viral infection and promoting viral survival. Moreover, during the course of MERS-CoV infection, the virus invades the immune system and down regulates MHC-I, MHC-II, and CD80/86 in antigen-presenting cells (APCs), which subsequently inhibit the T cell response [48]. These events may further impair the functions of B cells [49] and T cells via down regulation of the DPP4 receptors [29]. Recently, the induction of immunosuppression during the course of MERS-CoV infection, through promoting apoptosis of T cells, was identified as another strategy to manipulate survival pathways by the host immune response [50]. It has been thought that DPP4 may play significant roles in the signaling and activation of T cells during the course of MERS-COV infection [51]. Both $CD4^+$ and $CD8^+$ T cells isolated from human peripheral blood (PB), tonsils, spleens, and lymphoid organs could be infected with MERS-CoV but not with SARS-CoV. This infection pattern might be attributed to the low expression of the SARS-CoV receptor, namely, angiotensin-converting enzyme 2 (ACE2), in T cells [49]. A recent study reported that $CD4^+$ helper T cells were more susceptible to MERS-CoV infection. Additionally, MERS-CoV can induce T cell apoptosis by activating both the intrinsic and extrinsic apoptosis pathways [49]. Interestingly, there was a significant upregulation in the expression level of IL17 in MERS-CoV-infected patients [39]. T helper cells, especially Th17 cells, produce the proinflammatory cytokine IL17 via the STAT3 and NF-κB signaling pathways [52]. This finding suggests that MERS-CoV infection promotes the induction of Th17 cytokines. These Th17 cytokines can recruit neutrophils and monocytes to the site of infection or inflammation and lead to the activation of other downstream cytokine and chemokine cascades, such as IL1, IL6, TNF-α, TGF-β, IL8, and MCP-1 [53].

MERS-CoV-Antibody-Mediated Response

Neutralizing antibodies are very potent in neutralizing viral infectivity through blocking their entry into host cells. The detection of specific antibodies to MERS-CoV in human serum is considered one of the confirmative diagnoses for infection with MERS-CoV. It is crucial to determine whether these antibodies are potent arms of the adaptive response to MERS-CoV infection. However, detection of the anti-MERS-CoV antibody response occurs on days 14–21 after infection [54–56]. The antibody concentrations increase with time and last more than 18 months, and the long-term antibody response depends on the severity of the infection [57]. The anti-SARS-CoV antibody response can remain detectable for up to 24 months postinfection

[58] and then begins to gradually decrease until it completely disappears 6 years after infection [59].It is known that coronaviruses express surface spike glycoproteins, which are considered predominant antigenic proteins that stimulate the antibody response [60]. These antibodies might be used for targeting spike proteins and blocking the entry of the virus into host cells [60]. Therefore, designing monoclonal antibodies directed against these proteins is preferable for protection in contrast with vaccine preparation, which is a time-consuming and laborious process. To date, no vaccine for either SARS-CoV or MERS-CoV is available in the market despite some laboratory clinical trials.

A study reported by Coleman et al. proved that mice vaccinated with coronavirus S nanoparticle technology generate a high level of neutralizing antibodies against homologous viruses. These antibodies are not cross-protective with heterologous viruses [60]. On the other hand, a previous study performed by Chan et al. showed that antibodies recovered from the serum of some convalescent SARS patients might cross-react with MERS-CoV or neutralize it [61]. Another study suspected that these antibodies may cross-react with MERS-CoV because the epitope that is recognized by the cross-reactive antibodies might not be situated in the Spike 1 protein of SARS-CoV or at least might not be present in the receptor binding domain (RBD) [62].The finding of anti-MERS-CoV antibodies in Kenya in 1992 [63] was consistent with the results reported recently from Saudi Arabia. These findings suggest that MERS-CoV has been circulating in dromedary camels for more than 20 years in Saudi Arabia [64]. A total of 52.2% of these antibodies that were specific to the spike protein required a high titer to neutralize MERS-CoV, with a range between 1 : 80 to 1 : 800, and only 6% had a neutralizing antibody titer (more than 1 : 800) [63].

By using a recombinant MERS-CoV spike protein subunit 1-based ELISA (rELISA) [65], it was found that the antibody against spike protein was optimal for screening. It was also determined that 29.5% of serum samples isolated from dromedary camels were positive when tested by rELISA. In addition, all positive samples were tested using an established recombinant immunofluorescence assay, which showed that 93.4% of the samples had antibodies against MERS-CoV [63].

By using an anti-MERS-CoV nucleocapsid indirect ELISA and following 34 months of infection with MERS-CoV, the neutralizing antibody titers at 34 months of infection in 86% of human serum samples were the same as those after 13 months of infection. However, 29% of patients had a lowered titer of neutralizing antibody after 34 months of infection [66]. The low titer of antibodies in a few patients is attributed to viral shedding and persistence at the site of infection (i.e., mucosal site) or might be attributed to the neutralizing effect, which leads to a reduction in the proportion of antibodies. On the other hand, the long-term persistence of antibodies in most patients might be explained by the MERS-CoV infection inducing long-lived memory B cells, which in turn form antibody-secreting plasma cells that are stored in the bone morrow until reexposure to the same virus or similar epitopes. Thus, these antibodies may protect humans from reinfection with MERS-CoV, even though the concentration of antibodies in the serum is low. Thus, we can conclude that the type of assay used should be reconsidered to have a great sensitivity for viral detection. Further testing is required to identify conserved proteins in this virus serotype to induce effective antibody-mediated immunity as well as cell-mediated immunity.

In fact, a longitudinal study carried out in dromedary camels in the United Arab Emirates (UAE) between 2014 and 2015 demonstrated that serum samples collected on the day of dam parturition had a high level of specific anti-MERS-CoV antibodies, but in calves, the antibodies were not detectable [67]. The reason for this result might be that the camel calves consumed very low amounts of colostrum during the first 24 hours and that the low levels of IgG antibodies in the dams' milk started to decline at 24 hours postparturition [68]. However, the level of specific anti-MERS-CoV antibodies in the serum from calves increased, peaked on day 7 postparturition, and then decreased during the next 6 months; thus, neutralizing activity was functionally lost in 50% of the calves, and the rest had low antibody titers [67]. This result may explain the survival of MERS-CoV in these calves, and serum IgG antibodies may not be sufficient for protective immunity. Despite potential neutralizing antibodies in the serum, anti-MERS-CoV antibodies might be used as a valuable indication for viral diagnosis. Thus, a serum sample that is positive for specific antibodies would serve as a confirmative diagnosis of MERS-CoV infection.

MERS-COV VACCINES

In this section, we summarize the most recent findings with respect to MERS-CoV vaccine development, particularly vector- and RBD-based vaccines.

MERS-CoV Viral Vector-Based Vaccine

Recombinant-modified vaccinia virus Ankara (MVA) expressing the full-length MERS-CoV spike protein induced a high-level specific neutralizing antibody response in vaccinated BALB/c mice via intramuscular (i.m.) [69, 70] or subcutaneous (s.c.) routes of injection [70]. Several studies have reported that the i.m., s.c., and intradermal routes (i.d.) used for vaccine administration provide a good level of protection against both MERS-CoV and SARS-CoV infections. This protection level may be attributed to the downstream stimulation of a favorable immune response [71, 72]. However, this type of vaccine may elicit antibody-mediated disease enhancement (ADE) by the nonneutralizing epitopes encoded by the S glycoprotein [73]. MVA-MERS-CoV-S-specific neutralizing antibody titers were highly detectable after either a single immunization (day 21) or booster immunization (day 40) with a dose of 10^7 or 10^8 plaque forming units (PFUs) [70]. Compared to the antibody levels in previous studies of SARS-CoV [74], these high antibody levels were efficient in blocking the epitopes of MERS-CoV spike protein. The confirmation of antibody specificity was carried out by testing the serum obtained after the second booster against SARS-CoV; however, the serum showed undetectable levels of neutralizing antibody to SARS-CoV [69]. Previously, a correlation between the levels of specific neutralizing antibodies to the spike protein and the protectiveness of immunization in animals infected with SARS-CoV was shown [74, 75]. Thus, MVA-MERS-S vaccination can effectively stimulate humoral and cell-mediated responses. Additionally, the vaccine efficiency was similar in a study that conducted clinical testing, especially with respect to the immunogenicity of other recombinant MVA vaccines [76, 77]. MVA-MERS-CoV-S elicited a specific IFN-γ-producing $CD8^+$ T cell response against MERS-CoV infection by both the i.m. and s.c. routes following a prime-and-boost immunization regime. The specific $CD8^+$ T cells from the mouse spleen were stimulated with the MERS-CoV-S291 peptide and showed an upregulation in IFN-γ expression. The booster vaccination increased the level of the MERS-S291-specific $CD8^+$ T cell response by 3-fold [70]. Another study using an adenovirus type 5- (Ad5-) based vaccine expressing MERS-CoV-S proteins demonstrated the ability of this vaccine to induce systemic and mucosal antigen-specific immunity when administered via the i.m. or intragastric (i.g.) route [78]. This study proved that the sera of the vaccinated mice had high levels of antigen-specific IgGs and neutralizing antibodies, but no specific T cell response was detected in the case of the vaccines administered through the i.g. route. However, immunization through the i.m. route generated persistent antigen-specific T cell responses in both the spleen and lungs of the vaccinated animals [78]. Protective neutralizing antibodies and T cell-mediated responses were strongly elicited after challenging monkeys immunized with an adenoviral-based SARS vaccine expressing S1, M, and NP proteins with SARS-CoV [79]. Additionally, the elicitation of both humoral and cell-mediated responses has been confirmed with an adenoviral-based SARS-CoV vaccine encoding RBD [80].On the other hand, the preexisting immune response against the MVA and adenovirus vectors is one of the limitations of using viral vector-based vaccines, which may cause harmful immune responses and inflammation [69, 81, 82]. Although the MVA-MERS-CoV-S vector is a strong inducer of both cellular and antibody responses, there are some concerns about the safety of using these vector-based vaccines.

MERS-CoV-RBD-Based Vaccine

Some studies have shown that the RBD-based subunit of the SARS-CoV vaccine is very effective and safer than the viral vector candidates [83, 84]. As reported, immunization of mice with an RBD-based vaccine by the i.m. route induces long-term protection against SARS-CoV infection [85]. Thus, targeting MERS-CoV-RBD protein-1 is one of the strategies for vaccine development [86]. The immunogenicity of this fragment within the MERS-CoV-S spike protein was tested and evaluated. Remarkably, the MERS-CoV-S377-588-Fc has stronger immunogenicity than the other MERS-CoV-RBD proteins (S367-388-Fc, S358-588-Fc, and S367-606-Fc) and elicits significantly higher titers of neutralizing antibodies in vaccinated mice [87]. These antibodies are capable of blocking the binding of MERS-CoV-RBD to its receptors. This is a promising trend in the development of effective and safe MERS-CoV vaccines [88, 89]. Two available antibodies (REGN3051 and REGN3048) were capable of binding the RBD of the S protein and inhibiting its interaction with DPP4. Therefore, a potential inhibitor was developed [90]. These antibodies were tested in a mouse model and were at least effective at inhibiting MERS-CoV replication [91], but further testing of these vaccine candidates in dromedary camels should be conducted. Moreover, a developed humanized monoclonal antibody (mAb YS110) against DPP4 was reported and demonstrated inhibition of MERS-CoV infections [91]. Another study revealed that both mice and rabbits develop high titers of neutralizing antibodies when stimulated with 377-588-Fc [87]. Intranasal (i.n.) vaccination with a MERS-CoV-RBD-based subunit vaccine has a strong potential to induce a mucosal neutralizing IgA response against the RBD and MERS-CoV S proteins [87].

The 358-588 RBD fragment was shown to induce neutralizing antibodies in immunized rabbits [89], whereas fragment 377-662 was effective in immunized mice [88]. These results demonstrated that the expression of the recombinant S377-662-Fc protein in the RBD vaccine potentially triggers the production of specific antibodies in mice through the s.c. route after two booster vaccinations; these neutralizing antibodies are effective against MERS-CoV in Vero E6 cells in vitro [88]. Because this pathogen is mucosal-dependent, the administration of an i.n. vaccine that stimulates a potent mucosal IgA response would be a better route for the induction of an increased mucosal immune response to prevent infection with MERS-CoV. It has been reported that both local and systemic immunity are induced effectively by the i.n. immunization pathway. The mucosal IgA in vaccinated animals can provide cross-protection against homologous and heterologous strains of influenza virus and lead to long-term protection due to the memory response [92]. Moreover, long-term protection was reported in SARS-CoV infection by the i.n. route [80]. Zhang et al. reported some promising results using S377-662-Fc protein for mouse immunization via the i.n. route [93]. In contrast to the s.c. pathway, mice immunized intranasally with S377-662-Fc protein exhibited markedly high levels of specific IgA in the lungs [93]. However, compared with the sera from mice treated by the s.c. route, the sera of mice immunized through the i.n. route contained slightly higher levels of mucosal IgA [94]. On the other hand, the levels of IgG in mice immunized with S377-662-Fc protein via both s.c. and i.n. routes were the same. After a single dose administered via the i.n. route, the level of IgG was relatively low [94]. After several doses over 6 months, the titers of MERS-CoV-S1-specific IgG were high and persisted long term. These data confirm that the MERS-CoV-S377-662-Fc protein is capable of triggering a strong local mucosal response, especially by the i.n. route [93]. Moreover, a specific cell-mediated immune response in the spleen of immunized mice was generated by MERS-CoV-S377-662-Fc protein [94]. This finding indicates that both mucosal humoral and cellular immune responses might contribute to MERS-CoV prevention once induced by the RBD subunit vaccine. Due to the ability of the MERS-CoV-RBD-based vaccine to induce effective systematic and mucosal neutralizing antibodies, this subunit might be considered a promising potential candidate for the prevention of MERS-CoV infection [93]. A strong specific IgG antibody response against RBD was generated by Ad-MERS-S 4 to 16 weeks postimmunization. The levels of induced antibodies were significantly higher after immunization via the i.m. route than after immunization via the i.g. route. This result implies that this recombinant virus is capable of inducing a long-term specific antibody response via both routes [78].

The Potential Uses of Adjuvants in Association with MERS-CoV Vaccines

The neutralizing antibody production caused by most coronaviruses usually increases when immunization is used in conjunction with an adjuvant. It was reported that the inoculation of mice with MERS-CoV spike protein alone did not induce sufficient antibody production unless the viral protein was bound to an adjuvant, which then caused a potent response of neutralizing antibodies [60]. Both alum and MF59 adjuvants can elicit antigen-specific antibodies and cellular-mediated responses [95] and might be used for MERS-CoV subunit vaccine administration. However, alum alone cannot induce a potent Th1 response unless combined with another adjuvant, such as glucopyranosyl lipid A (a synthetic TLR-4 agonist). This cocktail will improve the effectiveness of the MERS-CoV-RBD-based subunit vaccines [96]. Coleman et al. showed that the immune responses to both SARS-CoV and MERS-CoV-S nanoparticles were increased significantly by approximately 15- and 7-fold by using the adjuvants alum and MF59, respectively [60]. This study was consistent with another investigation in terms of the production of anti-SARS-CoV neutralizing antibodies in mice [97]. Matrix M1 consists of two different components of saponin fractions: Matrix-A (Fraction-C saponin), which is the weaker part of saponin, and Matrix-C (Fraction-C saponin), which is a highly active adjuvant [98]. In clinical trials, Matrix M1 was proven to be a potent adjuvant [99]. In another study, using Matrix M1 as an adjuvant significantly boosted the level of antibody titers by 68-fold. [60]. The utilization of adjuvants might enhance immunogenicity and safety in MERS-CoV vaccine development.

FUTURE PROSPECTS

The innate immune response is an important element of antiviral defense and adaptive immunity. Further investigation is required to achieve a better understanding of the innate immune response to MERS-CoV. Thus, having sufficient data on highly pathogenic MERS-CoV, including understanding the mediators of innate immunity, their pathways, and how this virus can be regulated, will pave the way to develop effective antiviral therapeutics and vaccine candidates. To provoke a specific immune response without disease progression, an effective vaccine should be formulated. To date, there is no effective vaccine targeting the specific protein antigens of MERS-CoV.

MERS-CoV contains other accessory viral proteins (M, ORF 4a, ORF 4b, and ORF 5) [100]. In addition to the MERS-CoV-S protein, the membrane (matrix) protein and other structural proteins may have important roles in the development of other vaccine candidates. Thus, targeting these viral proteins might facilitate vaccine development [73] by limiting their ability to inhibit IFN production through binding to the dsRNA of the virus [100–102]. Additionally, the MERS-CoV-RBD-based vaccine, particularly the specific residues 377-606, induced strong and high antibody titers that were shown to have a neutralizing effect against MERS-CoV infection in immunized animals. However, testing these residues in human cells is required to confirm their efficacy as a potential vaccine. Thus, a protein-based vaccine that contains RBD should have great potential to elicit a highly neutralizing antibody response against several epitopes [73]. Additionally, as MERS-CoV targets the mucosa of the respiratory tract, designing a vaccine that enhances the induction of strong immunity via the i.n. route would be one of the best strategies to block MERS-CoV infection. Several factors affect the immunogenicity of vaccines. Each vaccine has an appropriate route of administration. Thus, selection of the optimal route of administration and the proper adjuvant with a specific, conserved antigen will play significant roles in MERS-CoV vaccine development and the efficacy of these candidate vaccines.

CONFLICTS OF INTEREST

The authors have no conflicts of interest to declare.

ACKNOWLEDGMENTS

The authors would like to thank the Deanship of Scientific Research at Majmaah University for supporting this work under project No. 1440-1. This grant is mainly dedicated to authors affiliated with Majmaah University.

REFERENCES

1. D. Butler, "Clusters of coronavirus cases put scientists on alert," Nature, vol. 492, no. 7428, pp. 166-167, 2012.
2. World Health Organization, Countries Agree Next Steps to Combat Global Health Threat by MERS-CoV, WHO, 2019.
3. J. F. W. Chan, S. K. P. Lau, K. K. W. To, V. C. C. Cheng, P. C. Y. Woo, and K. Y. Yuen, "Middle East respiratory syndrome coronavirus: another zoonotic betacoronavirus causing SARS-like disease," Clinical Microbiology Reviews, vol. 28, no. 2, pp. 465–522, 2015.
4. A. M. Zaki, S. van Boheemen, T. M. Bestebroer, A. D. M. E. Osterhaus, and R. A. M. Fouchier, "Isolation of a novel coronavirus from a man with pneumonia in Saudi Arabia," The New England Journal of Medicine, vol. 367, no. 19, pp. 1814–1820, 2012.
5. I. Eckerle, V. M. Corman, M. A. Muller, M. Lenk, R. G. Ulrich, and C. Drosten, "Replicative capacity of MERS coronavirus in livestock cell lines," Emerging Infectious Diseases, vol. 20, no. 2, pp. 276–279, 2014.
6. Z. A. Memish, N. Mishra, K. J. Olival et al., "Middle East respiratory syndrome coronavirus in bats, Saudi Arabia," Emerging Infectious Diseases, vol. 19, no. 11, pp. 1819–1823, 2013.
7. A. Assiri, G. R. Abedi, A. A. B. Saeed et al., "Multifacility outbreak of Middle East respiratory syndrome in Taif, Saudi Arabia," Emerging Infectious Diseases, vol. 22, no. 1, pp. 32–40, 2016.
8. A. J. Alsahafi and A. C. Cheng, "The epidemiology of Middle East respiratory syndrome coronavirus in the Kingdom of Saudi Arabia, 2012–2015," International Journal of Infectious Diseases, vol. 45, pp. 1–4, 2016.
9. R. Breban, J. Riou, and A. Fontanet, "Interhuman transmissibility of Middle East respiratory syndrome coronavirus: estimation of pandemic risk," The Lancet, vol. 382, no. 9893, pp. 694–699, 2013.
10. M. Cotten, S. J. Watson, P. Kellam et al., "Transmission and evolution of the Middle East respiratory syndrome coronavirus in Saudi Arabia: a descriptive genomic study," The Lancet, vol. 382, no. 9909, pp. 1993–2002, 2013.
11. S. Xia, Q. Liu, Q. Wang et al., "Middle East respiratory syndrome coronavirus (MERS-CoV) entry inhibitors targeting spike protein," Virus Research, vol. 194, pp. 200–210, 2014.

12. G. Lu, Y. Hu, Q. Wang et al., "Molecular basis of binding between novel human coronavirus MERS-CoV and its receptor CD26," Nature, vol. 500, no. 7461, pp. 227–231, 2013.
13. A. A. Al-Qahtani, K. Lyroni, M. Aznaourova et al., "Middle east respiratory syndrome corona virus spike glycoprotein suppresses macrophage responses via DPP4-mediated induction of IRAK-M and PPARγ," Oncotarget, vol. 8, no. 6, pp. 9053–9066, 2017.
14. H. J. Crespo, J. T. Y. Lau, and P. A. Videira, "Dendritic cells: a spot on sialic acid," Frontiers in Immunology, vol. 4, pp. 491–491, 2013.
15. S. Balachandran, P. C. Roberts, L. E. Brown et al., "Essential role for the dsRNA-dependent protein kinase PKR in innate immunity to viral infection," Immunity, vol. 13, no. 1, pp. 129–141, 2000.
16. H. Kato, K. Takahasi, and T. Fujita, "RIG-I-like receptors: cytoplasmic sensors for non-self RNA," Immunological Reviews, vol. 243, no. 1, pp. 91–98, 2011.
17. M. A. Anwar, S. Basith, and S. Choi, "Negative regulatory approaches to the attenuation of Toll-like receptor signaling," Experimental & Molecular Medicine, vol. 45, no. 2, article e11, 2013.
18. J. Krishnan, K. Selvarajoo, M. Tsuchiya, G. Lee, and S. Choi, "Toll-like receptor signal transduction," Experimental & Molecular Medicine, vol. 39, no. 4, pp. 421–438, 2007.
19. H. Kumar, T. Kawai, and S. Akira, "Toll-like receptors and innate immunity," Biochemical and Biophysical Research Communications, vol. 388, no. 4, pp. 621–625, 2009.
20. K. Honda and T. Taniguchi, "IRFs: master regulators of signalling by Toll-like receptors and cytosolic pattern-recognition receptors," Nature Reviews. Immunology, vol. 6, no. 9, pp. 644–658, 2006.
21. J. Brown, H. Wang, G. N. Hajishengallis, and M. Martin, "TLR-signaling networks: an integration of adaptor molecules, kinases, and cross-talk," Journal of Dental Research, vol. 90, no. 4, pp. 417–427, 2011.
22. L. A. J. O'Neill and A. G. Bowie, "The family of five: TIR-domain-containing adaptors in Toll-like receptor signalling," Nature Reviews Immunology, vol. 7, no. 5, pp. 353–364, 2007.
23. A. L. Totura, A. Whitmore, S. Agnihothram et al., "Toll-like receptor 3 signaling via TRIF contributes to a protective innate immune response to severe acute respiratory syndrome coronavirus infection," mBio,

vol. 6, no. 3, article e00638, 2015.

24. T. Sheahan, T. E. Morrison, W. Funkhouser et al., "MyD88 is required for protection from lethal infection with a mouse-adapted SARS-CoV," PLoS Pathogens, vol. 4, no. 12, article e1000240, 2008.
25. M. Yoneyama, M. Kikuchi, T. Natsukawa et al., "The RNA helicase RIG-I has an essential function in double-stranded RNA-induced innate antiviral responses," Nature Immunology, vol. 5, no. 7, pp. 730–737, 2004.
26. R. M. Welsh and S. N. Waggoner, "NK cells controlling virus-specific T cells: rheostats for acute vs. persistent infections," Virology, vol. 435, no. 1, pp. 37–45, 2013.
27. T. Taniguchi and A. Takaoka, "A weak signal for strong responses: interferon-alpha/beta revisited," Nature Reviews Molecular Cell Biology, vol. 2, no. 5, pp. 378–386, 2001.
28. J. Zhou, H. Chu, C. Li et al., "Active replication of Middle East respiratory syndrome coronavirus and aberrant induction of inflammatory cytokines and chemokines in human macrophages: implications for pathogenesis," The Journal of Infectious Diseases, vol. 209, no. 9, pp. 1331–1342, 2014.
29. H. Chu, J. Zhou, B. Ho-Yin Wong et al., "Productive replication of Middle East respiratory syndrome coronavirus in monocyte-derived dendritic cells modulates innate immune response," Virology, vol. 454–455, pp. 197–205, 2014.
30. R. W. Y. Chan, M. C. W. Chan, S. Agnihothram et al., "Tropism of and innate immune responses to the novel human betacoronavirus lineage C virus in human ex vivo respiratory organ cultures," Journal of Virology, vol. 87, no. 12, pp. 6604–6614, 2013.
31. F. Zielecki, M. Weber, M. Eickmann et al., "Human cell tropism and innate immune system interactions of human respiratory coronavirus EMC compared to those of severe acute respiratory syndrome coronavirus," Journal of Virology, vol. 87, no. 9, pp. 5300–5304, 2013.
32. J. Canton, A. R. Fehr, R. Fernandez-Delgado et al., "MERS-CoV 4b protein interferes with the NF-κB-dependent innate immune response during infection," PLoS Pathogens, vol. 14, no. 1, p. e1006838, 2018.
33. A. H. de Wilde, V. S. Raj, D. Oudshoorn et al., "MERS-coronavirus replication induces severe in vitro cytopathology and is strongly inhibited by cyclosporin A or interferon-α treatment," The Journal of

General Virology, vol. 94, Part 8, pp. 1749–1760, 2013.
34. B. J. Hart, J. Dyall, E. Postnikova et al., "Interferon-β and mycophenolic acid are potent inhibitors of Middle East respiratory syndrome coronavirus in cell-based assays," The Journal of General Virology, vol. 95, Part 3, pp. 571–577, 2014.
35. Q. Pan, P. E. de Ruiter, H. J. Metselaar et al., "Mycophenolic acid augments interferon-stimulated gene expression and inhibits hepatitis C virus infection in vitro and in vivo," Hepatology, vol. 55, no. 6, pp. 1673–1683, 2012.
36. J. M. Nicholls, L. L. M. Poon, K. C. Lee et al., "Lung pathology of fatal severe acute respiratory syndrome," The Lancet, vol. 361, no. 9371, pp. 1773–1778, 2003.
37. E. Faure, J. Poissy, A. Goffard et al., "Distinct immune response in two MERS-CoV-infected patients: can we go from bench to bedside?" PLoS One, vol. 9, no. 2, article e88716, 2014.
38. J. Zhou, H. Chu, J. F.-W. Chan, and K.-Y. Yuen, "Middle East respiratory syndrome coronavirus infection: virus-host cell interactions and implications on pathogenesis," Virology Journal, vol. 12, no. 1, pp. 218–218, 2015.
39. W. H. Mahallawi, O. F. Khabour, Q. Zhang, H. M. Makhdoum, and B. A. Suliman, "MERS-CoV infection in humans is associated with a pro-inflammatory Th1 and Th17 cytokine profile," Cytokine, vol. 104, pp. 8–13, 2018.
40. M. J. Cameron, L. Ran, L. Xu et al., "Interferon-mediated immunopathological events are associated with atypical innate and adaptive immune responses in patients with severe acute respiratory syndrome," Journal of Virology, vol. 81, no. 16, pp. 8692–8706, 2007.
41. L. Dumoutier and J. C. Renauld, "Viral and cellular interleukin-10 (IL-10)-related cytokines: from structures to functions," European Cytokine Network, vol. 13, no. 1, pp. 5–15, 2002.
42. R. Sabat, "IL-10 family of cytokines," Cytokine & Growth Factor Reviews, vol. 21, no. 5, pp. 315–324, 2010.
43. S. D. Blackburn and E. J. Wherry, "IL-10, T cell exhaustion and viral persistence," Trends in Microbiology, vol. 15, no. 4, pp. 143–146, 2007.
44. X. Xu and X. Gao, "Immunological responses against SARS-coronavirus infection in humans," Cellular & Molecular Immunology,

vol. 1, no. 2, pp. 119–122, 2004.

45. J. Zhao, K. Li, C. Wohlford-Lenane et al., "Rapid generation of a mouse model for Middle East respiratory syndrome," Proceedings of the National Academy of Sciences of the United States of America, vol. 111, no. 13, pp. 4970–4975, 2014.
46. Z. Y. Yang, W. P. Kong, Y. Huang et al., "A DNA vaccine induces SARS coronavirus neutralization and protective immunity in mice," Nature, vol. 428, no. 6982, pp. 561–564, 2004.
47. R. Channappanavar, J. Zhao, and S. Perlman, "T cell-mediated immune response to respiratory coronaviruses," Immunologic Research, vol. 59, no. 1-3, pp. 118–128, 2014.
48. L. Josset, V. D. Menachery, L. E. Gralinski et al., "Cell host response to infection with novel human coronavirus EMC predicts potential antivirals and important differences with SARS coronavirus," MBio, vol. 4, no. 3, 2013.
49. T. Ying, W. Li, and D. S. Dimitrov, "Discovery of T-cell infection and apoptosis by Middle East respiratory syndrome coronavirus," The Journal of Infectious Diseases, vol. 213, no. 6, pp. 877–879, 2016.
50. H. Chu, J. Zhou, B. H. Y. Wong et al., "Middle East respiratory syndrome coronavirus efficiently infects human primary T lymphocytes and activates the extrinsic and intrinsic apoptosis pathways," The Journal of Infectious Diseases, vol. 213, no. 6, pp. 904–914, 2016.
51. T. Ishii, K. Ohnuma, A. Murakami et al., "CD26-mediated signaling for T cell activation occurs in lipid rafts through its association with CD45RO," Proceedings of the National Academy of Sciences of the United States of America, vol. 98, no. 21, pp. 12138–12143, 2001.
52. M. L. Manni, K. M. Robinson, and J. F. Alcorn, "A tale of two cytokines: IL-17 and IL-22 in asthma and infection," Expert Review of Respiratory Medicine, vol. 8, no. 1, pp. 25–42, 2013.
53. W. Jin and C. Dong, "IL-17 cytokines in immunity and inflammation," Emerging Microbes & Infections, vol. 2, no. 1, pp. 1–5, 2013.
54. V. M. Corman, A. M. Albarrak, A. S. Omrani et al., "Viral shedding and antibody response in 37 patients with Middle East respiratory syndrome coronavirus infection," Clinical Infectious Diseases, vol. 62, no. 4, pp. 477–483, 2016.
55. W. B. Park, R. A. P. M. Perera, P. G. Choe et al., "Kinetics of serologic responses to MERS coronavirus infection in humans, South Korea,"

Emerging Infectious Diseases, vol. 21, no. 12, pp. 2186–2189, 2015.

56. U. Buchholz, M. A. Müller, A. Nitsche et al., "Contact investigation of a case of human novel coronavirus infection treated in a German hospital, October–November 2012," Euro Surveillance, vol. 18, 2013.
57. A. N. Alshukairi, I. Khalid, W. A. Ahmed et al., "Antibody response and disease severity in healthcare worker MERS survivors," Emerging Infectious Diseases, vol. 22, no. 6, 2016.
58. W. Liu, A. Fontanet, P. H. Zhang et al., "Two-year prospective study of the humoral immune response of patients with severe acute respiratory syndrome," The Journal of Infectious Diseases, vol. 193, no. 6, pp. 792–795, 2006.
59. F. Tang, Y. Quan, Z.-T. Xin et al., "Lack of peripheral memory B cell responses in recovered patients with severe acute respiratory syndrome: a six-year follow-up study," The Journal of Immunology, vol. 186, no. 12, pp. 7264–7268, 2011.
60. C. M. Coleman, Y. V. Liu, H. Mu et al., "Purified coronavirus spike protein nanoparticles induce coronavirus neutralizing antibodies in mice," Vaccine, vol. 32, no. 26, pp. 3169–3174, 2014.
61. K. H. Chan, J. F. W. Chan, H. Tse et al., "Cross-reactive antibodies in convalescent SARS patients' sera against the emerging novel human coronavirus EMC (2012) by both immunofluorescent and neutralizing antibody tests," Journal of Infection, vol. 67, no. 2, pp. 130–140, 2013.
62. L. Du, C. Ma, and S. Jiang, "Antibodies induced by receptor-binding domain in spike protein of SARS-CoV do not cross-neutralize the novel human coronavirus hCoV-EMC," The Journal of Infection, vol. 67, no. 4, pp. 348–350, 2013.
63. V. M. Corman, J. Jores, B. Meyer et al., "Antibodies against MERS coronavirus in dromedary camels, Kenya, 1992–2013," Emerging Infectious Diseases, vol. 20, no. 8, pp. 1319–1322, 2014.
64. A. N. Alagaili, T. Briese, N. Mishra et al., "Middle East respiratory syndrome coronavirus infection in dromedary camels in Saudi Arabia," MBio, vol. 5, no. 2, 2014.
65. Z. A. Memish, M. Cotten, B. Meyer et al., "Human infection with MERS coronavirus after exposure to infected camels, Saudi Arabia, 2013," Emerging Infectious Diseases, vol. 20, no. 6, pp. 1012–1015, 2014.
66. D. C. Payne, I. Iblan, B. Rha et al., "Persistence of antibodies against

Middle East respiratory syndrome coronavirus," Emerging Infectious Diseases, vol. 22, no. 10, pp. 1824–1826, 2016.

67. B. Meyer, J. Juhasz, R. Barua et al., "Time course of MERS-CoV infection and immunity in dromedary camels," Emerging Infectious Diseases, vol. 22, no. 12, pp. 2171–2173, 2016.

68. R. Kamber, Z. Farah, P. Rusch, and M. Hassig, "Studies on the supply of immunoglobulin G to newborn camel calves (Camelus dromedarius)," The Journal of Dairy Research, vol. 68, no. 1, pp. 1–7, 2001.

69. F. Song, R. Fux, L. B. Provacia et al., "Middle East respiratory syndrome coronavirus spike protein delivered by modified vaccinia virus Ankara efficiently induces virus-neutralizing antibodies," Journal of Virology, vol. 87, no. 21, pp. 11950–11954, 2013.

70. A. Volz, A. Kupke, F. Song et al., "Protective efficacy of recombinant modified vaccinia virus Ankara delivering Middle East respiratory syndrome coronavirus spike glycoprotein," Journal of Virology, vol. 89, no. 16, pp. 8651–8656, 2015.

71. L. Du, Z. Kou, C. Ma et al., "A truncated receptor-binding domain of MERS-CoV spike protein potently inhibits MERS-CoV infection and induces strong neutralizing antibody responses: implication for developing therapeutics and vaccines," PLoS One, vol. 8, no. 12, article e81587, 2013.

72. N. Takasuka, H. Fujii, Y. Takahashi et al., "A subcutaneously injected UV-inactivated SARS coronavirus vaccine elicits systemic humoral immunity in mice," International Immunology, vol. 16, no. 10, pp. 1423–1430, 2004.

73. N. Zhang, S. Jiang, and L. Du, "Current advancements and potential strategies in the development of MERS-CoV vaccines," Expert Review of Vaccines, vol. 13, no. 6, pp. 761–774, 2014.

74. H. Bisht, A. Roberts, L. Vogel et al., "Severe acute respiratory syndrome coronavirus spike protein expressed by attenuated vaccinia virus protectively immunizes mice," Proceedings of the National Academy of Sciences of the United States of America, vol. 101, no. 17, pp. 6641–6646, 2004.

75. Z. Chen, L. Zhang, C. Qin et al., "Recombinant modified vaccinia virus Ankara expressing the spike glycoprotein of severe acute respiratory syndrome coronavirus induces protective neutralizing antibodies primarily targeting the receptor binding region," Journal of Virology,

vol. 79, no. 5, pp. 2678–2688, 2005.

76. J. H. C. M. Kreijtz, M. Goeijenbier, F. M. Moesker et al., "Safety and immunogenicity of a modified-vaccinia-virus-Ankara-based influenza A H5N1 vaccine: a randomised, double-blind phase 1/2a clinical trial," The Lancet Infectious Diseases, vol. 14, no. 12, pp. 1196–1207, 2014.
77. A. F. Altenburg, J. H. C. M. Kreijtz, R. D. De Vries et al., "Modified vaccinia virus Ankara (MVA) as production platform for vaccines against influenza and other viral respiratory diseases," Viruses, vol. 6, no. 7, pp. 2735–2761, 2014.
78. X. Guo, Y. Deng, H. Chen et al., "Systemic and mucosal immunity in mice elicited by a single immunization with human adenovirus type 5 or 41 vector-based vaccines carrying the spike protein of Middle East respiratory syndrome coronavirus," Immunology, vol. 145, no. 4, pp. 476–484, 2015.
79. W. Gao, A. Tamin, A. Soloff et al., "Effects of a SARS-associated coronavirus vaccine in monkeys," The Lancet, vol. 362, no. 9399, pp. 1895-1896, 2003.
80. L. Du, G. Zhao, Y. Lin et al., "Intranasal vaccination of recombinant adeno-associated virus encoding receptor-binding domain of severe acute respiratory syndrome coronavirus (SARS-CoV) spike protein induces strong mucosal immune responses and provides long-term protection against SARS-CoV infection," Journal of Immunology, vol. 180, no. 2, pp. 948–956, 2008.
81. S. L. Pichla-Gollon, S. W. Lin, S. E. Hensley et al., "Effect of preexisting immunity on an adenovirus vaccine vector: in vitro neutralization assays fail to predict inhibition by antiviral antibody in vivo," Journal of Virology, vol. 83, no. 11, pp. 5567–5573, 2009.
82. H. Weingartl, M. Czub, S. Czub et al., "Immunization with modified vaccinia virus Ankara-based recombinant vaccine against severe acute respiratory syndrome is associated with enhanced hepatitis in ferrets," Journal of Virology, vol. 78, no. 22, pp. 12672–12676, 2004.
83. L. Du, Y. He, Y. Zhou, S. Liu, B. J. Zheng, and S. Jiang, "The spike protein of SARS-CoV—a target for vaccine and therapeutic development," Nature Reviews. Microbiology, vol. 7, no. 3, pp. 226–236, 2009.
84. Y. He, H. Lu, P. Siddiqui, Y. Zhou, and S. Jiang, "Receptor-binding domain of severe acute respiratory syndrome coronavirus spike protein contains multiple conformation-dependent epitopes that induce highly

potent neutralizing antibodies," Journal of Immunology, vol. 174, no. 8, pp. 4908–4915, 2005.

85. L. Du, G. Zhao, Y. He et al., "Receptor-binding domain of SARS-CoV spike protein induces long-term protective immunity in an animal model," Vaccine, vol. 25, no. 15, pp. 2832–2838, 2007.
86. J. Tang, N. Zhang, X. Tao et al., "Optimization of antigen dose for a receptor-binding domain-based subunit vaccine against MERS coronavirus," Human Vaccines & Immunotherapeutics, vol. 11, no. 5, pp. 1244–1250, 2015.
87. C. Ma, L. Wang, X. Tao et al., "Searching for an ideal vaccine candidate among different MERS coronavirus receptor-binding fragments—the importance of immunofocusing in subunit vaccine design," Vaccine, vol. 32, no. 46, pp. 6170–6176, 2014.
88. L. Du, G. Zhao, Z. Kou et al., "Identification of a receptor-binding domain in the S protein of the novel human coronavirus Middle East respiratory syndrome coronavirus as an essential target for vaccine development," Journal of Virology, vol. 87, no. 17, pp. 9939–9942, 2013.
89. H. Mou, V. S. Raj, F. J. M. van Kuppeveld, P. J. M. Rottier, B. L. Haagmans, and B. J. Bosch, "The receptor binding domain of the new Middle East respiratory syndrome coronavirus maps to a 231-residue region in the spike protein that efficiently elicits neutralizing antibodies," Journal of Virology, vol. 87, no. 16, pp. 9379–9383, 2013.
90. K. E. Pascal, C. M. Coleman, A. O. Mujica et al., "Pre- and postexposure efficacy of fully human antibodies against Spike protein in a novel humanized mouse model of MERS-CoV infection," Proceedings of the National Academy of Sciences of the United States of America, vol. 112, no. 28, pp. 8738–8743, 2015.
91. A. S. Agrawal, T. Garron, X. Tao et al., "Generation of a transgenic mouse model of Middle East respiratory syndrome coronavirus infection and disease," Journal of Virology, vol. 89, no. 7, pp. 3659–3670, 2015.
92. F. Y. Liew, S. M. Russell, G. Appleyard, C. M. Brand, and J. Beale, "Cross-protection in mice infected with influenza A virus by the respiratory route is correlated with local IgA antibody rather than serum antibody or cytotoxic T cell reactivity," European Journal of Immunology, vol. 14, no. 4, pp. 350–356, 1984.
93. N. Zhang, J. Tang, L. Lu, S. Jiang, and L. Du, "Receptor-binding domain-based subunit vaccines against MERS-CoV," Virus Research,

vol. 202, pp. 151–159, 2015.

94. C. Ma, Y. Li, L. Wang et al., "Intranasal vaccination with recombinant receptor-binding domain of MERS-CoV spike protein induces much stronger local mucosal immune responses than subcutaneous immunization: implication for designing novel mucosal MERS vaccines," Vaccine, vol. 32, no. 18, pp. 2100–2108, 2014.
95. A. Podda and G. Del Giudice, "MF59-adjuvanted vaccines: increased immunogenicity with an optimal safety profile," Expert Review of Vaccines, vol. 2, no. 2, pp. 197–204, 2003.
96. S. Jiang, M. E. Bottazzi, L. Du et al., "Roadmap to developing a recombinant coronavirus S protein receptor-binding domain vaccine for severe acute respiratory syndrome," Expert Review of Vaccines, vol. 11, no. 12, pp. 1405–1413, 2012.
97. Y. V. Liu, M. J. Massare, D. L. Barnard et al., "Chimeric severe acute respiratory syndrome coronavirus (SARS-CoV) S glycoprotein and influenza matrix 1 efficiently form virus-like particles (VLPs) that protect mice against challenge with SARS-CoV," Vaccine, vol. 29, no. 38, pp. 6606–6613, 2011.
98. K. Lovgren Bengtsson, B. Morein, and A. D. Osterhaus, "ISCOM technology-based Matrix M™ adjuvant: success in future vaccines relies on formulation," Expert Review of Vaccines, vol. 10, no. 4, pp. 401–403, 2011.
99. R. J. Cox, G. Pedersen, A. S. Madhun et al., "Evaluation of a virosomal H5N1 vaccine formulated with Matrix M™ adjuvant in a phase I clinical trial," Vaccine, vol. 29, no. 45, pp. 8049–8059, 2011.
100. Y. Yang, L. Zhang, H. Geng et al., "The structural and accessory proteins M, ORF 4a, ORF 4b, and ORF 5 of Middle East respiratory syndrome coronavirus (MERS-CoV) are potent interferon antagonists," Protein & Cell, vol. 4, no. 12, pp. 951–961, 2013.
101. K. L. Siu, M. L. Yeung, K. H. Kok et al., "Middle east respiratory syndrome coronavirus 4a protein is a double-stranded RNA-binding protein that suppresses PACT-induced activation of RIG-I and MDA5 in the innate antiviral response," Journal of Virology, vol. 88, no. 9, pp. 4866–4876, 2014.
102. D. Niemeyer, T. Zillinger, D. Muth et al., "Middle East respiratory syndrome coronavirus accessory protein 4a is a type I interferon antagonist," Journal of Virology, vol. 87, no. 22, pp. 12489–12495, 2013.

12

Evolution of Knowledge, Awareness, and Practices regarding Zika Virus from 2016 to 2017

Quinton Katler,[1] Prachi Godiwala,[1] Charles Macri,[1] Beth Pineles,[2] Gary Simon,[3] Aileen Chang,[3] and Homa Ahmadzia[1]

[1]Department of Obstetrics and Gynecology, The George Washington University Hospital, Washington, DC, USA

[2]Department of Obstetrics, Gynecology and Reproductive Sciences, University of Maryland Medical Center, Baltimore, MD, USA

[3]Department of Internal Medicine, The George Washington University Hospital, Washington, DC, USA

ABSTRACT

Objective. Our team created a knowledge, attitudes, and practice (KAP) survey in order to assess changes over time in healthcare provider and community member awareness of Zika virus symptoms, transmission, treatment, and current and future concerns. *Study Design*. The cross-sectional survey was issued at an academic medical center in Washington,

Citation: Evolution of Knowledge, Awareness, and Practices regarding Zika Virus from 2016 to 2017. Quinton Katler, Prachi Godiwala, Charles Macri, Beth Pineles, Gary Simon, Aileen Chang, and Homa Ahmadzia. Hindawi Infectious Diseases in Obstetrics and Gynecology Volume 2017, Article ID 6350602, 13 pages https://doi.org/10.1155/2017/6350602

DC, and via an online link to healthcare providers and community members between June and August 2016. Survey distribution was then repeated the following year, from March to April 2017. Outcomes were compared by survey year and healthcare provider versus community member status using SAS Program Version 9.4. *Results*. Significant differences in knowledge, attitudes, and practices existed between 2016 and 2017 survey time points. By 2017, more respondents had knowledge of various Zika virus infection characteristics; however healthcare provider knowledge also waned in certain areas. Attitudes towards Zika virus infection displayed an overall decreased concern by 2017. Practice trends by 2017 demonstrated fewer travel restrictions to Zika-endemic areas and increased mosquito protective measures within the US. *Conclusions*. Our results provide novel insight into the transformation of knowledge, attitudes, and practice of community members and healthcare providers regarding Zika virus since its declaration as a public health emergency of international concern in 2016.

INTRODUCTION

Zika virus infection was declared a public health emergency in 2016. Zika virus infections have been present in Africa and Asia since the 1940s and have since spread dramatically to mainly in the Americas and western Pacific [1]. Zika virus was first identified in a rhesus monkey in Uganda in 1947, during which time the virus vector was discovered as the *Aedes africanus* mosquito [2–4]. Since the initial detection of Zika virus, epidemiologists have mapped its global migration through the remainder of the 20th century [5]. In October of 2015, Brazil reported "an unusual increase" in the number of cases of microcephaly among newborns. After increased investigative efforts, Brazil reported detection of Zika virus in amniotic fluid samples from pregnant women with ultrasound-confirmed microcephaly, culminating in the issuance of a national public health emergency [6]. Zika virus reached the United States (US) in mid-2016 when the first local-transmission was documented in Miami on July 29, 2016 [7]. As of July 2017, there had been over 143 symptomatic Zika virus disease cases reported within the US and over 513 cases within US Territories [8].

Since the first documented cases in the western hemisphere, the scientific community has made substantial progress in terms of understanding Zika virus transmission and outcomes. It is known that an infected mother can pass the Zika virus vertically to her fetus during pregnancy as well as during the delivery of her infant [9]. The Zika virus can also be passed through sexual intercourse, even if the infected person is asymptomatic [9]. There are reports of Zika virus infections acquired through laboratory exposure [9]. Other modes of transmission are being investigated, such as blood transfusion and healthcare occupational exposure. The most common symptoms of a Zika infection include fever, rash, headache, joint pain, conjunctivitis, and muscle pain. Zika virus has been associated with specific patterns of neurologic and anatomic abnormalities in newborns, including microcephaly (congenital small head size leading to brain abnormalities), decreased brain tissue volume, and ocular damage [9]. Zika infection is diagnosed by screening of symptoms and recent travel history followed by testing of blood or urine. While there is no commercially available vaccine or targeted treatment for Zika virus, symptoms are treated by rest, hydration, and fever-reducing medication [9].

Despite substantial progress in understanding Zika virus pathophysiology among the scientific community, to our knowledge there is no study or data regarding the evolution in knowledge, awareness, and practice among community members and healthcare providers with regard to Zika virus during this time period. Given this void, the aim of our study was to analyze the change in knowledge, awareness, and prevention/practice regarding Zika infection during a critical time in the development of understanding of the virus.

STUDY DESIGN

A Zika questionnaire was composed by a multidisciplinary team of experts consisting of maternal-fetal-medicine physicians, public health specialists, general medicine providers, and obstetrician-gynecologist generalist physicians. This cross-sectional survey was distributed in two platforms. First, patients and other community members visiting the outpatient obstetrics and gynecology clinics at a mid-size academic center in Washington, DC, were approached by medical student study volunteers to take the survey prior to their scheduled appointments over a 6-week period from late June to early August 2016. This time period was chosen because, during this interval, information about Zika virus and practices to safely prevent transmission were at the forefront of media attention due to many uncertainties surrounding the virus, and it is a common time for our study participants to be traveling to areas prevalent with Zika virus. The survey was loaded onto handheld tablet computers via the secure RED Cap platform and volunteers approached respondents at random during normal clinic hours; the intervals of office survey administration were dependent on the availability of volunteers. Second, the survey was emailed to community members and healthcare providers using RED Cap to collect the data. The RED Cap survey link was emailed to various listservs within the academic medical community, including OB/GYN resident and faculty listservs, medical student listservs, and the Rodham Institute. The survey was also emailed to nonmedical personnel via social media networks in order to obtain a random sampling of community members. The survey distribution was then repeated in the same fashion over 6 weeks nearly one year later, from early March to late April 2017. This time period was chosen because the authors felt that sufficient time had passed since the first survey administration to gauge a change in knowledge, attitudes, and practices; Zika virus garnered much less media attention around this time due to increased availability of information regarding the virus.Healthcare providers were defined as anyone involved in the medical field and were stratified as either a physician (primary care, OB/GYN, or other specialty physicians), or nonphysician, which could include but was not limited to medical students, NPs, PAs, genetic counselors, and other types of providers. Information about the type of no physician healthcare providers was not collected. Community members were defined as anyone receiving care at the clinic where the survey was administered or anyone whose occupation was not in the healthcare field. The survey sample size included 174 respondents (63% community members; 37% healthcare providers) during the first round and 277 respondents (65% community members; 35% healthcare providers) during the second round. No face-to-face interviews were conducted. At the end of the survey the respondent was provided with a link to the most current CDC Guidelines on Zika virus.

Data was stored in the George Washington University REDCap database. Outcomes were compared by year of survey administration and healthcare provider versus community member status. Data analysis was performed using SAS Program Version 9.4. The median and interquartile range was computed for continuous variables. Frequencies were computed for categorical variables. Chi-square tests were used to compare frequencies of categorical variables across groups. p values < 0.05 were considered statistically significant.

The study was approved by the Office of Human Research-Institutional Review Board at The George Washington University (Protocol #031653, approved April 20, 2016). Respondent identity was anonymous to protect confidentiality.

RESULTS

Of the approximately 800 individuals approached to take the survey either in the clinic or via email link in each round, approximately half were healthcare providers or half were community members, and 451 potential participants responded. Approximately 75 of these were community members approached in the waiting room of the clinic. Of the 451 total survey respondents, there were more respondents in the second round of surveys (277) compared to the first (174). Fifty-seven respondents (33%) in the first round of surveys were healthcare providers and 60% of these were physicians, whereas 89 (32%) were healthcare providers in the second round but only 37% of these were physicians. The population surveyed was majority Caucasian females with private medical insurance (Table 1).

Table 1. Demographic characteristics of survey respondents.

Character-istic	Details	2016 Survey Frequency (%)	2017 Survey Frequency (%)
Gender	Male	28/174 (16)	40/277 (14)
	Female	129/174 (74)	215/277 (78)
	No answer	17/174 (10)	22/277 (8)
Marital status	Single	52/174 (30)	125/277 (45)
	Partnered	18/174 (10)	33/277 (12)
	Married	79/174 (45)	89/277 (32)
	Other	25/174 (15)	30/277 (11)
Highest level of education	High School	12/174 (7)	0/277 (0)
	College	32/174 (18)	84/277 (30)
	Graduate school	60/174 (34)	112/277 (40)
	Post-Grad	53/174 (30)	60/277 (22)
	No answer	17/174 (10)	21/277 (8)
Average income	<$25,000	36/174 (21)	90/277 (32)
	$25,000–50,000	12/174 (7)	30/277 (11)
	$50,000–75,000	31/174 (18)	44/277 (16)
	$75,000–100,000	17/174 (10)	24/277 (9)
	>$100,000	39/174 (22)	39/277 (14)
	Prefer not to answer	17/174 (10)	29/277 (10)
	No answer	22/174 (13)	21/277 (8)

Ethnicity	Caucasian	92/174 (53)	188/277 (68)
	African American	27/174 (15)	14/277 (5)
	Hispanic	11/174 (6)	11/277 (4)
	Asian	18/174 (10)	31/277 (11)
	Other	9/174 (5)	12/277 (4)
	No answer	17/174 (10)	21/277 (8)
County of birth	US	136/174 (78)	219/277 (79)
	Other	21/174 (12)	36/277 (13)
	No answer	17/174 (10)	22/277 (8)
Country of residency	US	156/174 (90)	252/277 (91)
	Other	0/174 (0)	2/277 (1)
	No answer	18/174 (10)	23/277 (8)
State of residence	Washington, DC	67/174 (39)	73/277 (26)
	Maryland	30/174 (17)	32/277 (12)
	Virginia	29/174 (17)	52/277 (19)
	Other state	10/174 (6)	60/277 (22)
	No answer	38/174 (22)	60/277 (22)
Type of insurance	Medicare	8/174 (5)	2/277 (1)
	Medicaid	11/174 (6)	4/277 (1)
	Private	134/174 (77)	244/277 (88)
	None	1/174 (1)	4/277 (1)
	No answer	20/174 (11)	23/277 (8)
Healthcare provider status	Primary Care physician	10/174 (6)	8/277 (3)
	OB/GYN physician	16/174 (9)	15/277 (5)
	Other physician	8/174 (5)	10/277 (4)
	Non-physician provider	23/174 (13)	56/277 (20)
	Non-provider	117/174 (67)	188/277 (68)
Pregnancy status	Not pregnant	93/174 (53)	205/277 (74)
	Pregnant	32/174 (18)	10/277 (4)
	Planning to be pregnant	8/174 (5)	9/277 (3)
	Postpartum (<6 wk)	3/174 (2)	1/277 (0)
	Not applicable	21/174 (12)	31/277 (11)
	No answer	17/174 (10)	21/277 (8)

Primary source of healthcare information	Health publication	30/174 (17)	37/277 (13)
	Internet	55/174 (32)	100/277 (36)
	Social Media	2/174 (1)	4/277 (1)
	Healthcare provider	36/174 (21)	56/277 (20)
	Newspaper	6/174 (3)	12/277 (4)
	At work	7/174 (4)	21/277 (8)
	Other	20/174 (11)	26/277 (9)
	No answer	18/174 (10)	21/277 (8)
Where would you like to obtain your healthcare information	Text Message	7/174 (4)	8/277 (3)
	Health publication	31/174 (18)	43/277 (16)
	Internet	45/174 (26)	90/277 (32)
	Social Media	4/174 (2)	15/277 (5)
	Healthcare provider	43/174 (25)	61/277 (22)
	Newspaper	8/174 (5)	9/277 (3)
	Other	19/174 (11)	30/277 (11)
	No answer	17/174 (10)	21/277 (8)

Significant differences in knowledge existed between 2016 and 2017 survey time points (Table 2). More community members were aware of Zika virus in the second round of surveys (99% in 2017 versus 94% in 2016, $p = 0.02$). More respondents were able to correctly identify Zika infection symptoms in 2017, and this remained true for both community members (64% versus 46%, $p = 0.019$) as well as healthcare providers (79% versus 74%, $p = 0.19$). When asked if anyone had been infected with Zika virus within the US at the time of survey collection, more community members were aware in 2017 compared to 2016 (91% versus 65%, $p \leq 0.0001$), and a similar trend was noted for healthcare providers (84% versus 61%, $p = 0.002$). Knowledge with regard to a link between Zika infection during pregnancy and microcephaly increased among community respondents (95% in 2017 versus 90% in 2016, $p = 0.09$), and this knowledge remained basically unchanged among healthcare providers (100% versus 98%, $p = 0.25$). Knowledge about safety of mosquito-bite prevention strategies such as DEET and permethrin during pregnancy actually decreased over time, including for both healthcare providers and community members with a lower proportion of responses indicating knowledge of DEET and permethrin safety in pregnancy in 2017. However, a substantial number of respondents answered "I do not know" to this question in both years. Knowledge regarding effective protective measures against mosquitos changed among community members over time: in 2017, 93% (versus 82% in 2016) would remove standing water ($p = 0.015$), 96% (versus 78%) would sleep under a mosquito net ($p \leq 0.0001$), and 96% (versus 85%) would use screens on windows and doors ($p = 0.0049$); healthcare provider knowledge essentially remained the same between the survey time points.

Table 2. Responses to "knowledge" questions regarding Zika virus disease and infection.

	Community Member Responses			Healthcare Provider Responses		
	2016 Survey (%)	2017 Survey (%)	95% CI	2016 Survey (%)	2017 Survey (%)	95% CI
Are you aware of Zika virus?						
Yes	91/97 (94)	165/166 (99)	0.02	57/57 (100)	87/89 (98)	0.25
No	2/97 (2)	0/166 (0)		0/57 (0)	2/89 (2)	
No answer	4/97 (4)	1/166 (1)		0/57 (0)	0/89 (0)	
How is the Zika virus spread?						
Fly bite	0/97 (0)	1/166 (1)		0/57 (0)	0/89 (0)	
Mosquito Bite	92/97 (95)	163/166 (98)	0.19	56/57 (98)	87/89 (98)	0.24
Do not know	1/97 (1)	1/166 (1)		1/57 (2)	0/89 (0)	
No answer	4/97 (4)	1/166 (1)		0/57 (0)	2/89 (2)	
Which of the following is NOT a common symptom of Zika virus infection?						
Fever	2/97 (2)	1/166 (1)		0/57 (0)	2/89 (2)	
Rash	3/97 (3)	5/166 (3)		4/57 (7)	2/89 (2)	
Pneumonia	45/97 (46)	107/166 (64)	0.019	42/57 (74)	70/89 (79)	0.19
Joint Pain	2/97 (2)	2/166 (1)		1/57 (2)	0/89 (0)	
Conjunctivitis (red eyes)	15/97 (15)	28/166 (17)		5/57 (9)	3/89 (3)	
Do not know	26/97 (27)	22/166 (13)		5/57 (9)	12/89 (13)	
No answer	4/97 (4)	1/166 (1)		0/57 (0)	2/89 (2)	
What percentage of people infected with Zika virus will have symptoms?						
20%	38/97 (39)	76/166 (46)		29/57 (51)	47/89 (53)	
40%	16/97 (16)	28/166 (17)	0.24	11/57 (19)	11/89 (12)	0.52
60%	3/97 (3)	7/166 (4)		2/57 (4)	1/89 (1)	
80%	1/97 (1)	1/166 (1)		0/57 (0)	1/89 (1)	

Do not know	34/97 (35)	53/166 (32)		15/57 (26)	27/89 (30)	
No answer	5/97 (5)	1/166 (1)		0/57 (0)	2/89 (2)	
What is the treatment for Zika virus infection?						
Antibiotics	5/97 (5)	10/166 (6)		0/57 (0)	2/89 (2)	
Antivirals	26/97 (27)	53/166 (32)		0/57 (0)	3/89 (3)	
Antifungals	0/97 (0)	0/166 (0)		7/57 (12)	19/89 (21)	
Tylenol and rest	21/97 (22)	59/166 (36)	0.009	41/57 (72)	50/89 (56)	0.18
Do not know	41/97 (42)	43/166 (26)		9/57 (16)	15/89 (17)	
No answer	4/97 (4)	1/166 (1)		0/57 (0)	2/89 (2)	
Has anyone been infected with Zika virus in the US?						
Yes	63/97 (65)	151/166 (91)	0.0001	35/57 (61)	75/89 (84)	0.002
No	20/97 (21)	7/166 (4)		18/57 (32)	7/89 (8)	
Do not know	9/97 (9)	7/166 (4)		4/57 (7)	5/89 (5)	
No answer	5/97 (5)	1/166 (1)		0/57 (0)	2/89 (2)	
Has anyone been infected with Zika virus by traveling from the US to known Zika areas outside the US?						
Yes	84/97 (87)	151/166 (91)	0.24	53/57 (93)	83/89 (93)	0.64
No	4/97 (4)	6/166 (3)		1/57 (2)	1/89 (1)	
Do not know	5/97 (5)	8/166 (5)		3/57 (5)	3/89 (3)	
No answer	4/97 (4)	1/166 (1)		0/57 (0)	2/89 (2)	
Is there an approved vaccine for Zika virus?						
Yes	2/97 (2)	6/166 (4)		1/57 (2)	4/89 (4)	
No	74/97 (76)	144/166 (87)	0.048	56/57 (98)	75/89 (84)	0.048
Do not know	17/97 (17)	14/166 (8)		0/57 (0)	8/89 (9)	
No answer	4/97 (4)	2/166 (1)		0/57 (0)	2/89 (2)	
Is there a link between Zika virus and the birth defect microcephaly (small head)?						
Yes	87/97 (90)	157/166 (95)	0.09	57/57 (100)	87/89 (98)	0.25
No	0/97 (0)	1/166 (1)		0/57 (0)	0/89 (2)	
Do not know	5/97 (5)	7/166 (4)		0/57 (0)	0/89 (0)	
No answer	5/97 (5)	1/166 (1)		0/57 (0)	2/89 (2)	

Can an infected man transfer Zika virus to a woman through sexual activity?						
Yes	74/97 (76)	144/166 (87)	0.12	52/57 (91)	74/89 (83)	0.4
No	1/97 (1)	4/166 (2)		3/57 (5)	6/89 (7)	
Do not know	14/97 (14)	13/166 (8)		1/57 (2)	3/89 (3)	
It is unknown	3/97 (3)	2/166 (1)		1/57 (2)	1/89 (1)	
No answer	5/97 (5)	3/166 (2)		0/57 (0)	5/89 (5)	
Can an infected mother transfer Zika virus to her unborn baby?						
Yes	67/97 (69)	144/166 (87)	0.008	52/57 (91)	75/89 (84)	0.34
No	3/97 (3)	2/166 (1)		1/57 (2)	1/89 (1)	
Do not know	19/97 (19)	12/166 (7)		3/57 (5)	4/89 (5)	
It is unknown	3/97 (3)	5/166 (3)		1/57 (1)	3/89 (3)	
No answer	5/97 (5)	3/166 (2)		0/57 (0)	6/89 (6)	
What is the common breeding site for the mosquito that spreads Zika virus?						
standing water	78/97 (80)	155/166 (93)	0.007	55/57 (96)	80/89 (90)	0.38
running water	0/97 (0)	1/166 (1)		0/57 (0)	1/89 (1)	
salt water	0/97 (0)	1/166 (1)		0/57 (0)	0/89 (0)	
Do not know	13/97 (13)	6/166 (3)		2/57 (4)	5/89 (6)	
No answer	6/97 (6)	3/166 (2)		0/57 (0)	3/89 (3)	
Is DEET safe in pregnancy?						
Yes	24/97 (25)	34/166 (20)	0.19	28/57 (49)	30/89 (34)	0.05
No	29/97 (30)	60/166 (36)		19/57 (33)	26/89 (29)	
Do not know	38/97 (39)	69/166 (42)		10/57 (18)	29/89 (33)	
No answer	6/97 (6)	3/166 (2)		0/57 (0)	4/89 (4)	
Is permethrin safe in pregnancy?						
Yes	10/97 (10)	19/166 (11)	0.18	19/57 (33)	19/89 (21)	0.17
No	10/97 (10)	20/166 (12)		9/57 (16)	19/89 (21)	
Do not know	69/97 (71)	123/166 (74)		29/57 (51)	47/89 (53)	
No answer	8/97 (8)	4/166 (2)		0/57 (0)	4/89 (4)	

Where should you apply insect repellent?						
Under sunscreen	11/97 (11)	24/166 (14)		7/57 (12)	16/89 (18)	
Over sunscreen	43/97 (44)	87/166 (52)	0.05	38/57 (67)	38/89 (43)	0.04
Do not know	35/97 (36)	52/166 (31)		11/57 (19)	32/89 (36)	
No answer	8/97 (8)	3/166 (2)		1/57 (1)	3/89 (3)	
If infected with Zika virus, when is one most likely to infect others?						
First week of symptoms	20/97 (21)	51/166 (31)	0.09	26/57 (46)	29/89 (33)	0.24
Second week of symptoms	8/97 (8)	13/166 (8)		4/57 (7)	6/89 (6)	
Third week of symptoms	2/97 (2)	6/166 (4)		2/57 (4)	1/89 (1)	
Do not know	60/97 (62)	93/166 (56)		25/57 (44)	50/89 (56)	
No answer	7/97 (7)	3/166 (2)		1/57 (0)	3/89 (3)	
These precautions will help to reduce the risk of mosquito bites (yes/no)						
Remove standing water	80/97 (82)	154/166 (93)	0.015	56/57 (98)	81/89 (91)	0.29
Sleep under a mosquito net	76/97 (78)	160/166 (96)	0.0001	54/57 (95)	83/89 (93)	0.15
Use screens on windows and doors	82/97 (85)	160/166 (96)	0.0049	56/57 (98)	84/89 (94)	0.19

Community member knowledge appears to increase for certain topics while healthcare provider knowledge waned between the two time periods. Knowledge of Zika infection symptomatic treatment improved for the community respondents with 36% in 2017 being aware that Tylenol and rest was the primary modality of treatment compared to 22% in 2016 ($p = 0.009$).

However, this value trended differently for healthcare providers: 72% in 2016 had knowledge about treatment options compared to 56% in 2017 ($p = 0.18$).

More community members were able to correctly answer that a Zika vaccine was not commercially available when comparing 2017 to 2016 (87% versus 76%, $p = 0.048$); however the opposite trend was seen among healthcare providers (84% in 2017 versus 98% in 2016, $p = 0.048$). When questioned about the common breeding site of the Aedes mosquito, 93% of community members were able to correctly identify that the mosquito breeds near standing water compared to 80% in 2016 ($p = 0.007$); there was a small trend in the opposite direction for healthcare provider respondents (90% in 2017 versus 96% in 2016, $p = 0.38$).

Attitudes towards Zika infection changed over time (Table 3). The overall concern among community members for Zika infection being a serious problem in the respondents' home area decreased in 2017 survey administration (69% in 2017 versus 87% in 2016, $p \leq 0.0001$). This trend was similar for healthcare providers (80% versus 84%, $p = 0.18$).

When asked if Zika virus prevalence makes them worried for their families, less community members (29% versus 35%, $p = 0.02$) and healthcare providers (24% versus 30%, $p = 0.18$) were concerned in 2017. Additionally, more community members believed that a vaccine against Zika virus infection was important when comparing 2017 to 2016 (89% versus 80%, $p = 0.095$); notably, this trend was not witnessed among healthcare providers (87% versus 89%, $p = 0.03$).

With regard to screening all pregnant patients for Zika exposure or infection, 61% of community members in both rounds of surveys believed that this should be a routine prenatal policy ($p = 0.0083$). Notably, more healthcare providers believed that all pregnant mothers should be screened when comparing 2017 to 2016 (53% versus 46%, $p = 0.085$). Overall concern among community members for Zika virus transmission through sexual activity remained unchanged; however a decline in concern was noted among healthcare providers (70% in 2017 versus 82% 2016, $p = 0.089$).

Approximately 36% of community members were not concerned about having a child with microcephaly secondary to a Zika infection, which was an improvement from 2016 (26%, $p = 0.01$); a similar trend was noted for healthcare provider respondents over this time period (33% in 2017 versus 46% in 2016, $p = 0.02$).

When asked about having a child with learning disabilities due to a Zika infection, less community members (26% in 2017 versus 37% in 2016, $p = 0.009$) and healthcare providers (34% in 2017 versus 40% in 2016, $p = 0.04$) were concerned by the second round of survey distribution.

Table 3. Responses to "attitude" questions regarding Zika virus disease and infection.

	Community Member Responses			Healthcare Provider Responses		
	2016 Survey (%)	2017 Survey (%)	95% CI	2016 Survey (%)	2017 Survey (%)	95% CI
Do you think Zika virus is a serious problem in your area?						
Yes	13/97 (13)	17/166 (10)		8/57 (14)	10/89 (11)	
No	67/97 (69)	144/166 (87)	0.0001	48/57 (84)	71/89 (80)	0.18
Do not know	9/97 (9)	0/166 (0)		1/57 (2)	1/89 (1)	
No answer	8/97 (8)	5/166 (3)		0/57 (0)	7/89 (7)	
Does the Zika virus make you worried for your family?						
Yes	34/97 (35)	48/166 (29)	0.02	17/57 (30)	21/89 (24)	0.18
No	50/97 (52)	109/166 (66)		36/57 (63)	61/89 (69)	
Do not know	5/97 (5)	1/166 (1)		2/57 (4)	0/89 (0)	
No answer	8/97 (8)	8/166 (5)		2/57 (4)	7/89 (7)	
Do you think that a vaccine against Zika virus is important?						
Yes	78/97 (80)	148/166 (89)	0.095	51/57 (89)	77/89 (87)	0.03
No	4/97 (4)	8/166 (5)		3/57 (5)	0/89 (0)	
Do not know	7/97 (7)	5/166 (3)		3/57 (5)	5/89 (6)	
No answer	8/97 (8)	5/166 (3)		0/57 (0)	7/89 (8)	
Should all newly pregnant mothers be screened for Zika?						
Yes	59/97 (61)	101/166 (61)	0.0083	26/57 (46)	47/89 (53)	0.085
No	28/97 (19)	52/166 (31)		27/57 (47)	30/89 (34)	
Do not know	12/97 (12)	8/166 (5)		4/57 (7)	5/89 (6)	
No answer	8/97 (8)	5/166 (3)		0/57 (0)	7/89 (8)	
Are you concerned about Zika virus spreading through sexual activity?						
Yes	65/97 (67)	120/166 (72)	0.18	47/57 (82)	62/89 (70)	0.089
No	17/97 (18)	34/166 (20)		9/57 (16)	15/89 (17)	
Do not know	7/97 (7)	7/166 (4)		1/57 (2)	5/89 (6)	
No answer	8/97 (8)	5/166 (3)		0/57 (0)	7/89 (8)	

Are you concerned about Zika spreading from mother to unborn baby?						
Yes	70/97 (72)	123/166 (74)		53/57 (93)	76/89 (86)	
No	9/97 (9)	30/166 (18)	0.018	2/57 (4)	4/89 (4)	0.17
Do not know	8/97 (8)	8/166 (5)		2/57 (4)	2/89 (2)	
No answer	10/97 (10)	5/166 (3)		0/57 (0)	7/89 (8)	
Are you concerned about having a child with microcephaly (small head) related to Zika infection?						
Yes	56/97 (58)	98/166 (59)		29/57 (51)	53/89 (60)	
No	25/97 (26)	60/166 (36)	0.01	26/57 (46)	29/89 (33)	0.02
Do not know	7/97 (7)	3/166 (2)		2/57 (4)	0/89 (0)	
No answer	9/97 (9)	5/166 (3)		0/57 (0)	7/89 (8)	
Are you concerned about having a child with learning disabilities related to Zika infection?						
Yes	54/97 (56)	95/166 (57)		30/57 (53)	51/89 (57)	
No	25/97 (26)	61/166 (37)	0.009	23/57 (40)	30/89 (34)	0.04
Do not know	9/97 (9)	4/166 (2)		3/57 (5)	0/89 (0)	
No answer	9/97 (9)	6/166 (4)		1/57 (2)	8/89 (9)	
If you went to an area where Zika is spread and are a man, are you concerned about spreading Zika sexually to your partner?						
Yes	10/97 (10)	21/166 (13)		13/57 (23)	21/89 (24)	
No	7/97 (7)	9/166 (5)	0.37	0/57 (0)	2/89 (2)	0.6
Do not know	3/97 (3)	5/166 (3)		2/57 (4)	2/89 (2)	
Not applicable	68/97 (70)	125/166 (75)		40/57 (70)	57/89 (64)	
No answer	9/97 (9)	6/166 (4)		2/57 (3)	7/89 (8)	

Practices related to Zika changed over time as well (Table 4). The overall trend in the second cohort compared to the first was that less community members planned to restrict travel to Zika-endemic areas (37% versus 49%, $p = 0.048$). A similar trend was noted among healthcare professionals, although not statistically significant (35% in 2017 versus 47% in 2016, $p = 0.19$). When asked if respondents would take protective measures against mosquito infections if traveling to a Zika-endemic area, there was no statistically significant change in planned practice from 2016 to 2017 for both community members and healthcare providers. Furthermore, the overall trend was noted by 2017 that both healthcare providers and community members alike would plan for greater mosquito preventive measures within the US, including greater use of insect repellent and sleeping under a mosquito net; there was no change in anticipated practice of using screens on windows or doors. When community members were asked if they were to go to an area where Zika virus is spread and were pregnant, less would refrain from intercourse during the pregnancy (5% in 2017 versus 10% in 2016, $p = 0.13$); a similar trend was seen if their partners traveled to a Zika-endemic area (8% versus 14%, $p = 0.13$). When healthcare providers responded to these same questions, there was no identifiable trend or statistically significant association. Although statistically insignificant, less community members plan to use condoms during the entire pregnancy if they were to go to an area where Zika virus is spread and are pregnant (11% in 2017 versus 4% in 2016, $p = 0.26$); a similar trend was noted if their partners went to a Zika-endemic area (8% versus 3%, $p = 0.31$). Again, when healthcare providers responded to these questions, there was no identifiable trend or statistically significant association.

Table 4. Responses to "practice" questions regarding Zika virus disease and infection.

	Community Member Responses			Healthcare Provider Responses		
	2016 Survey (%)	2017 Survey (%)	95% CI	2016 Survey (%)	2017 Survey (%)	95% CI
Do you plan to restrict your travel to a Zika endemic area?						
Yes	48/97 (49)	61/166 (37)	0.048	27/57 (47)	31/89 (35)	0.19
No	34/97 (35)	86/166 (52)		25/57 (44)	47/89 (53)	
Do not know	7/97 (7)	12/166 (7)		4/57 (7)	4/89 (5)	
No answer	8/97 (8)	7/166 (4)		1/57 (2)	7/89 (8)	
If you travel to a Zika endemic area, will you take protective measures to prevent a mosquito bite?						
Yes	85/97 (88)	147/166 (89)	0.11	53/57 (93)	79/89 (89)	0.09
No	0/97 (0)	7/166 (4)		3/57 (5)	3/89 (3)	
Do not know	4/97 (4)	3/166 (2)		1/57 (2)	0/89 (0)	
No answer	8/97 (8)	9/166 (5)		0/57 (0)	7/89 (8)	

Do you plan on taking protective measures against mosquito bites while in the US to prevent Zika infection?						
Yes	52/97 (54)	96/166 (58)	0.29	24/57 (42)	36/89 (41)	0.18
No	30/97 (31)	57/166 (34)		31/57 (54)	44/89 (50)	
Do not know	7/97 (7)	6/166 (4)		2/57 (4)	2/89 (2)	
No answer	8/97 (8)	7/166 (4)		0/57 (0)	7/89 (8)	
If you are in an area with mosquitos, you take the following preventive measures						
Use insect repellent	73/97 (75)	138/166 (83)	0.18	47/57 (83)	77/89 (87)	0.018
Remove standing water	62/97 (64)	115/166 (69)	0.56	43/57 (75)	61/69 (69)	0.15
Sleep under a mosquito net	29/97 (30)	60/166 (36)	0.12	11/57 (19)	31/89 (35)	0.003
Use screens on windows and doors	84/97 (87)	145/166 (87)	0.16	50/57 (88)	78/89 (88)	0.01
If you went to an area where Zika virus is spread and are pregnant, do you plan to not have sex during the entire pregnancy?						
Yes	10/97 (10)	9/166 (5)	0.13	5/57 (9)	7/89 (8)	0.24
No	8/97 (8)	30/166 (18)		10/57 (18)	12/89 (13)	
Do not know	19/97 (20)	24/166 (14)		7/57 (12)	12/89 (13)	
Not applicable	52/97 (54)	90/166 (54)		34/57 (60)	47/89 (53)	
No answer	8/97 (8)	13/166 (8)		1/57 (2)	11/89 (12)	
If you went to an area where Zika virus is spread and are pregnant, do you plan to use condoms during the entire pregnancy?						
Yes	18/97 (19)	29/166 (17)	0.26	13/57 (23)	14/89 (16)	0.2
No	4/97 (4)	18/166 (11)		4/57 (7)	6/89 (7)	
Do not know	16/97 (16)	17/166 (10)		4/57 (7)	10/89 (11)	
Not applicable	51/97 (53)	89/166 (54)		35/57 (61)	49/89 (55)	
No answer	8/97 (8)	13/166 (8)		1/57 (2)	10/89 (11)	

If your partner went to an area where Zika virus is spread and you are pregnant, do you plan to not have sex during the entire pregnancy?						
Yes	14/97 (14)	13/166 (8)	0.13	7/57 (12)	7/89 (8)	0.067
No	6/97 (6)	25/166 (15)		11/57 (19)	10/89 (11)	
Do not know	17/97 (18)	24/166 (14)		4/57 (7)	14/89 (16)	
Not applicable	52/97 (54)	89/166 (54)		34/57 (60)	48/89 (54)	
No answer	8/97 (8)	15/166 (9)		1/57 (2)	10/89 (11)	
If your partner went to an area where Zika virus is spread and you are pregnant, do you plan to use condoms during the entire pregnancy?						
Yes	19/97 (20)	30/166 (18)	0.31	14/57 (25)	16/89 (18)	0.26
No	3/97 (3)	14/166 (8)		4/57 (7)	5/89 (6)	
Do not know	16/97 (16)	17/166 (10)		4/57 (7)	8/89 (9)	
Not applicable	51/97 (53)	90/166 (54)		34/57 (60)	50/89 (56)	
No answer	8/97 (8)	15/166 (9)		1/57 (2)	10/89 (11)	
If you are thinking of becoming pregnant, do you plan to delay your pregnancy due to the Zika outbreak?						
Yes	11/97 (11)	20/166 (12)	0.78	10/57 (18)	17/89 (19)	0.23
No	12/97 (12)	30/166 (18)		8/57 (14)	9/89 (10)	
Do not know	11/97 (11)	16/166 (10)		4/57 (7)	5/89 (6)	
Not applicable	53/97 (55)	86/166 (52)		34/57 (60)	47/89 (53)	
No answer	10/97 (10)	14/166 (8)		1/57 (2)	11/89 (12)	

When comparing physician healthcare providers with nonphysician healthcare providers, notable differences occurred between the first and second rounds of survey administration. More physicians than no physicians were able to correctly identify symptoms of Zika infection in the first round of surveys (79% versus 65%, resp., $p = 0.002$) as well as correctly identify symptomatic treatment of Zika infection (85% versus 52%, $p = 0.001$). These knowledge gaps closed during the second round of surveys, with approximately the same amount of both physicians and nonphysician providers able to identify symptoms (76% and 80%, $p = 0.02$) and correct treatment (61% and 54%, $p = 0.007$). However, as can be noted from these data, more physicians were aware of the correct treatment for the virus in the first round (85%) than in the second round (61%). Interestingly, in the first round, fewer physicians than nonphysicians were aware that Zika virus can be transmitted sexually from a man to a woman (88% versus 96%, $p = 0.001$), which remained true in the second round (79% versus 86%, $p = 0.25$), although not statistically significant. Fewer physicians than nonphysicians were worried for their families due to Zika virus during the first round of surveys as well (21% versus 43%, $p = 0.02$), while the opposite was true in the second round (27% of physicians versus 21% of nonphysician providers, $p = 0.26$).

Similarly, fewer physicians than nonphysicians were worried about having a child with microcephaly related to Zika infection (41% versus 65%, $p = 0.0005$) or having a child with disabilities related to Zika infection (41% versus 70%, $p = 0.001$) in the first round. In the second round, however, these groups responded much more comparably; both physician and nonphysician providers alike were similarly worried about having a child with microcephaly (61% versus 59%, $p = 0.30$) or having a child with disabilities related to Zika infection (55% versus 59%, $p = 0.44$). As a general trend, the answers to most of the attitude and practice questions related to Zika virus and protection against it indicated that physicians were less concerned overall about the virus compared to nonphysician healthcare providers in both rounds of surveys, with the difference being less noticeable in the second round, but the majority of these were not statistically significant.

DISCUSSION

When comparing the community member results from the 2017 round of surveys to those of 2016, it is clear that there was an improvement over time in baseline knowledge with regard to Zika virus and infection. These respondents were more aware of Zika virus existence, virus vector, infectious symptoms, virus incubation period, and treatment options. Additionally, community members were more knowledgeable with regard to commercial vaccine availability, modes of transmission, mosquito breeding sites and prevention measures, and status of locally acquired Zika infections. More of these respondents were also aware of possible pregnancy and neonatal neurologic comorbidities resulting from a Zika infection. Weaker knowledge areas during both rounds of survey administration included the safety of mosquito prevention medications during pregnancy, including DEET and permethrin. DEET, the topical insect repellent, has been endorsed by the US Environmental Protection Agency to have low acute toxicity without appearing to pose a significant health concern to humans when used as directed [10]. Although some limited epidemiologic data suggests that DEET use in pregnancy may adversely affect learning and behavior, the evidence for this remains low [11]. Accordingly, permethrin is largely considered safe in pregnancy. While there are no randomized-controlled studies that distinguish outcomes based on human exposure during pregnancy, most retrospective studies find no adverse perinatal outcomes after permethrin use [12]. These results reveal that information regarding medication safety during pregnancy is not being translated to the general public appropriately, as most patients answered incorrectly or "I do not know" to these questions. Thus, increased counseling for healthcare providers is required on behalf of their patients. The overall improvement in knowledge among the community members that participated in our survey is promising, as this may parallel the general trend in Zika virus knowledge of general population as a whole. Since the majority of our survey respondents prefer to obtain their medical information from the Internet and healthcare providers (Table 1), this supports the notion that these resources have served as effective avenues for disseminating medical knowledge over the past year.

While there were notable improvements in healthcare provider knowledge during the survey periods, there were also areas where knowledge remained stagnant, or even decreased. More healthcare providers could properly identify Zika virus symptoms in 2017 compared to 2016, and there is high baseline knowledge among healthcare providers regarding mosquito prevention strategies as well as the links between virus infection and microcephaly. However, while the percentage of healthcare providers that were aware of locally acquired Zika infections within the US increased between survey rounds, this percentage was still less than 90% even in 2017. While the first documented cases of Zika infection within the US were from June to August 2016 [13] and the end of data collection for the first round of surveys was August 3, 2016, the second round of surveys occurred more than six months after the first local Zika virus infection was reported. Thus, increased healthcare provider awareness of the status of Zika virus infections within the US is warranted, as this percentage should ideally near 100%. Additionally, only about one-half of healthcare providers were able to report the percentage of infected individuals that will display symptoms. This gap in provider knowledge may lead to inadvertent patient harm including improper Zika screening, as well as inadequate patient counseling. Moreover, our survey discovered alarming areas with declining healthcare provider knowledge. Specifically, less healthcare providers were aware of DEET and permethrin safety in pregnancy in 2017 when compared to 2016, in parallel to a greater overall uncertainty with regard to medication safety as demonstrated by more "I do not know" responses. One reason for this may be that, as the initial media exposure about Zika virus subsided, new and novel information about the virus and its spread was less ubiquitous and therefore fewer providers were well informed during the 2017 survey administration. This decline in knowledge may have detrimental impact on patient counseling about safe virus prevention practices. Accordingly, there were similar declines in healthcare provider knowledge regarding options available for symptomatic treatment of Zika infections, awareness of a commercially available vaccine, and transmission of Zika virus vertically (from mother to fetus) and through sexual intercourse.

The results of the "attitude" questions between both survey periods reveal a tendency among both community members and healthcare providers to feel less concerned about Zika infection than they had previously. According to our results, there is decreased concern among both groups of survey respondents regarding concern for family members, risk of local infection, and transmission to unborn fetuses as well as postpartum neurologic outcomes for newborns. Interestingly, while the percentage of healthcare providers that were aware that Zika virus could be transferred through sexual intercourse or from mother to fetus decreased between surveys, there was a parallel decrease in concern for these outcomes among healthcare providers. The general trend of the public being less concerned about Zika virus likely derives from several ideas. First, the WHO lifted the declaration of Zika as a public health emergency of international concern, which may have contributed to alleviating fears among the general public [14]. Also, as more information is learned and distributed about Zika virus to healthcare professionals and the community alike, this likely helps to relieve anxiety and fear among the public. However, there are areas in our survey that limit our ability to fully extrapolate "attitude" information. For instance, the question "Are you concerned about having a child with learning disabilities due to a Zika infection?" does not clarify the timing of infection in relation to the pregnancy (i.e., infection before, during, or after pregnancy). Additionally, there was an upward shift among community members and healthcare providers alike in terms of belief in the need for universal Zika infection screening for pregnant women. However, we did not specify how we should screen (i.e., via questionnaire, by symptom or recent travel, or by serum testing) or when we should screen (prepregnancy versus trimester-specific screening); thus, respondents may have interpreted this question differently.

There are interesting trends to be noted when analyzing the results of the "practice" questions between both survey periods. The 2017 round of survey responses alludes to an overall decreased concern among community members with regard to planned travel to Zika-endemic areas, whether within the US or to an international site. However, it is important to review these results within the context of the demographics of the surveyed population. A majority (greater than 50%) of the survey respondents were not pregnant, nor were they planning on getting pregnant. Accordingly, there may have been an increased hesitancy for travel if a greater percentage of respondents were in a pregnancy-planning stage of their lives. Additionally, there appears to be a shift towards increased practice of mosquito prevention within the United States among community members and healthcare providers alike. Interestingly, this was not replicated when asked about mosquito prevention practices if planning to travel to Zika-endemic areas. As more confirmed cases of Zika infection occur within the United States, this may encourage the general public to begin to practice routine mosquito prevention practices, rather than only associated with travel.There are other limitations of our study that have been identified. One limitation is that the answers to some of the questions in the survey may have changed throughout survey administration. For example, one question asked the respondent whether any documented cases of Zika virus transmission in the US have occurred. In the first survey administration, the answer to this question changed as initially there were no documented cases, but towards the end of the survey administration period there had been several cases noted. This changes the distribution of answers and the ability to compare this to the second group. Additionally, when asked about vaccine availability, there may be potential confusion among healthcare provider respondents with regard to the current status of an "approved" Zika virus vaccine. At the time of survey administration, investigational vaccines were undergoing clinical trial [15] and this ambiguity may have contributed to the distribution of answers that were obtained for this question. Another limitation of our study is that the total number of respondents in the first group is significantly less than in the second group, so the ability to detect statistically significant differences within this group when comparing the first group to the second group is diminished. Accordingly, the cohort composition was different between the two time periods, which would limit our ability to measure change longitudinally. It is also unknown exactly how many potential participants the survey was sent to via email and how many were approached in clinic, as the number of participants on each listserv changes frequently and the number of participants approached in the clinic was not recorded. This discrepancy in the number of participants as well as potential differences in baseline characteristics of responders versus nonresponses to the survey may contribute an element of participation bias. However, it can be estimated that the number of potential participants was about 800 by analyzing current listserv cohorts and from an estimation of those approached in clinic. Another limitation is that the sampling of respondents was convenience sampling, meaning that the survey was distributed both in the clinic and via the online link based on the potential participants' ease of approachability and willingness to respond to the survey. It is important to note that our survey also did not stratify healthcare providers by their specific profession (i.e., ultrasound technician versus nurse versus other healthcare professionals) and only asked about physician versus nonphysician status; this may have played a role in skewing the survey results, as the second group of healthcare providers may not represent the same expertise or experience as the first group of surveyed providers.

CONCLUSION

Zika virus is a new consideration in the field of congenitally acquired diseases. This study aimed to address the evolving knowledge, attitudes, and practice regarding Zika virus during a tumultuous time of many unknowns for patients and providers alike. From this study, we can extrapolate that, as more information is gathered about the virus, more questions will be answered and there will be fewer fears regarding the virus. This data can be used to identify gaps in knowledge at this time about Zika virus and prepare medical providers to offer counseling regarding safety and preventive practices.

ACKNOWLEDGMENTS

The authors would like to thank the Department of Obstetrics and Gynecology at George Washington University Hospital, the MFA providers and staff, and the community for participating in their study. They would also like to thank Miriam Toaff, Zoe Belkin, Diana Shustarovich, Aislynn Raymond, and Anja Frost, for assisting with survey collection, and the members of the Rodham Institute for participating in the survey.

REFERENCES

1. N. Broutet, F. Krauer, M. Riesen et al., "Zika Virus as a Cause of Neurologic Disorders," *The New England Journal of Medicine*, vol. 374, no. 16, pp. 1506–1509, 2016.
2. G. W. A. Dick, S. F. Kitchen, and A. J. Haddow, "Zika Virus (I). Isolations and serological specificity," *Transactions of the Royal Society of Tropical Medicine and Hygiene*, vol. 46, no. 5, pp. 509–520, 1952.
3. G. Dick, "Zika virus (II). Pathogenicity and physical properties," *Transactions of the Royal Society of Tropical Medicine and Hygiene*, vol. 46, no. 5, pp. 521–534, 1952.
4. K. C. Smithburn, "Neutralizing antibodies against certain recently isolated viruses in the sera of human beings residing in East Africa," *The Journal of Immunology*, vol. 69, no. 2, pp. 223–34, 2017.
5. O. Faye, C. C. M. Freire, A. Iamarino et al., "Molecular evolution of Zika virus during its emergence in the 20(th) century," *PLOS Neglected Tropical Diseases*, vol. 8, no. 1, 2014.
6. World Health Organization, *The History of Zika Virus: The World Health Organization*, 2017.
7. Center for Disease Control and Prevention, *Press Release: Florida Investigation Links Four Recent Zika Cases to Local Mosquito-Borne Virus Transmission*, 2016.
8. Center for Disease Control and Prevention, *Zika Cases in the US – 2017*, 2017.
9. Center for Disease Control and Prevention, *Zika Virus: What we know 2017*, 2017.
10. B. J. Wylie, M. Hauptman, A. D. Woolf, and R. H. Goldman, "Insect Repellants During Pregnancy in the Era of the Zika Virus," *Obstetrics & Gynecology*, vol. 128, no. 5, pp. 1111–1115, 2016.
11. T. G. Osimitz and J. V. Murphy, "Neurological effects associated with use of the insect repellent *N,N*-diethyl-*m*-toluamide (DEET)," *Journal of Toxicology - Clinical Toxicology*, vol. 35, no. 5, pp. 435–441, 1997.
12. D. Kennedy, V. Hurst, E. Konradsdottir, and A. Einarson, "Pregnancy outcome following exposure to permethrin and use of teratogen information," *American Journal of Perinatology*, vol. 22, no. 2, pp. 87–90, 2005.

13. A. Likos, I. Griffin, A. M. Bingham et al., "Local Mosquito-Borne Transmission of Zika Virus — Miami-Dade and Broward Counties, Florida, June–August 2016," *Morbidity and Mortality Weekly Report (MMWR)*, vol. 65, no. 38, pp. 1032–1038, 2016.
14. M. Chan, "Zika: We must be ready for the long haul," in *World Health Organization Commentary., 2017., Available at Accessed*, 2017
15. National Institute of Allergy and Infectious Diseases, "News Release: Phase 2 Zika Vaccine Trial Begins in U.S., Central and South America," 2017, https://www.niaid.nih.gov/news-events/phase-2-zika-vaccine-trial-begins-us-central-and-south-america.

13

What Factors Might Have Led to the Emergence of Ebola in West Africa?

Kathleen A. Alexander,[1] Claire E. Sanderson,[1] Madav Marathe,[2,3] Bryan L. Lewis,[3] Caitlin M. Rivers,[3] Jeffrey Shaman,[4] John M. Drake,[5] Eric Lofgren,[3] Virginia M. Dato,[6] Marisa C. Eisenberg,[7] and Stephen Eubank[3]

[1] Department of Fisheries and Wildlife Conservation, Virginia Tech, Blacksburg, Virginia, United States of America,

[2] Department of Computer Science, Virginia Tech, Blacksburg, Virginia, United States of America,

[3] Network Dynamics and Simulation Science Laboratory, Virginia Bioinformatics Institute, Virginia Tech, Blacksburg, Virginia, United States of America,

[4] Department of Environmental Health Sciences, Mailman School of Public Health, Columbia University, New York, New York, United States of America,

[5] Odum School of Ecology, University of Georgia, Athens, Georgia, United States of America,

[6] Department of Biomedical Informatics, School of Medicine, University of Pittsburgh, Pittsburgh, Pennsylvania, United States of America,

[7] Departments of Epidemiology and Mathematics, University of Michigan, Ann Arbor, Michigan, United States of America,

Citation: What Factors Might Have Led to the Emergence of Ebola in West Africa? Alexander KA, Sanderson CE, Marathe M, Lewis BL, Rivers CM, Shaman J, et al. (2015). PLoS Negl Trop Dis 9(6): e0003652. doi:10.1371/journal.pntd.0003652

ABSTRACT

An Ebola outbreak of unprecedented scope emerged in West Africa in December 2013 and presently continues unabated in the countries of Guinea, Sierra Leone, and Liberia. Ebola is not new to Africa, and outbreaks have been confirmed as far back as 1976. The current West African Ebola outbreak is the largest ever recorded and differs dramatically from prior outbreaks in its duration, number of people affected, and geographic extent. The emergence of this deadly disease in West Africa invites many questions, foremost among these: why now, and why in West Africa? Here, we review the sociological, ecological, and environmental drivers that might have influenced the emergence of Ebola in this region of Africa and its spread throughout the region. Containment of the West African Ebola outbreak is the most pressing, immediate need. A comprehensive assessment of the drivers of Ebola emergence and sustained human-to-human transmission is also needed in order to prepare other countries for importation or emergence of this disease. Such assessment includes identification of country-level protocols and interagency policies for outbreak detection and rapid response, increased understanding of cultural and traditional risk factors within and between nations, delivery of culturally embedded public health education, and regional coordination and collaboration, particularly with governments and health ministries throughout Africa. Public health education is also urgently needed in countries outside of Africa in order to ensure that risk is properly understood and public concerns do not escalate unnecessarily. To prevent future outbreaks, coordinated, multiscale, early warning systems should be developed that make full use of these integrated assessments, partner with local communities in high-risk areas, and provide clearly defined response recommendations specific to the needs of each community.

INTRODUCTION

On December 6, 2013, the world's largest Ebola epidemic began when a two-year-old in Guéckédou, Guinea, a small village bordering Sierra Leone and Liberia, became infected (Fig 1) [1,2]. This is the first documented Ebola outbreak outside Central Africa and is unique in its size, duration, and spatial extent. The circulating virus has been identified as the Zaire ebolavirus (EBOV), a strain previously found in only three Central African countries: the Democratic Republic of the Congo (DRC), Republic of the Congo, and Gabon (Fig 1) [3]. The public health impact of the current Ebola epidemic in West Africa has been far greater than case counts. Massive indirect effects on already-weakened public services have occurred, including significant crippling of the health sector, which has increased the impacts of other endemic diseases and the associated mortality [3]. Substantial economic loss and social disruption will have a sustained impact on the region that will far outlive the actual epidemic [4]. One paper, published in 2011, argued that Ebola would never become a significant public health threat in Africa [4]; clearly, the threat of Ebola has been underestimated. The emergence of Ebola in West Africa invites many questions—most of which remain unresolved—notably: why now, and why in West Africa? Advancing our understanding of this outbreak remains critical to present health care interventions as well as the prevention of further outbreaks. Here, we review the sociological, ecological, and environmental drivers that could have influenced the emergence of EBOV in West Africa at this time and in this manner. Given these factors, we explore the lessons of this outbreak and evaluate how we might manage future threats from Ebola across the complex urban and rural landscapes that define modern Africa.

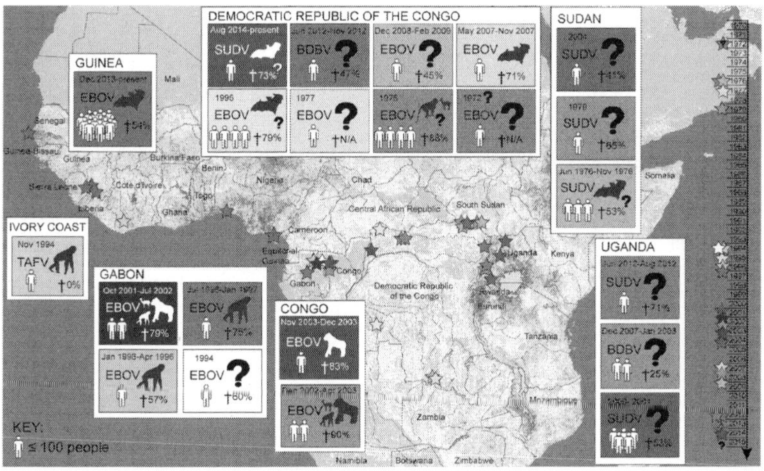

Figure 1. Map of Ebola outbreaks in Africa.

The outbreak in West Africa is unprecedented in its scope and duration, occurring for the first time in urban centers. Historically, Ebola viral outbreaks (stars, timeline right) occurred sporadically, limited largely to Central African rural areas where the human population (grey to red gradient stippling [5,6]) has been low or more remote from areas of high population density. It is uncertain how frequent Ebola outbreaks will be in the future, given the identification of wildlife spillover potential in West Africa and the increasingly concentrated human populations in this region.

EBOLA IN AFRICA

Ebola hemorrhagic fever is an emerging zoonotic viral disease that historically has occurred in rural areas of Central Africa, with isolated cases identified elsewhere (Fig 1 and Table 1). The Ebola virus was first identified in humans in southern Sudan in 1976 [7], but likely occurred as early as 1972 in Tandala, DRC [8]. The virus causes severe morbidity and high mortality in humans and wildlife [9]. Humans typically are infected with Ebola either through contact with bodily fluids of infected animals or humans, or through consumption of bush meat, caring for patients, or preparing the deceased for burial (Fig 2) [10]. EBOV can be found in a number of human secretions during the acute phase of infection, such as saliva, feces, semen, breast milk, tears, nasal blood, and skin [11]. Presently, there is no vaccine or other therapeutic interventions beyond supportive care, although promising pharmaceutical options are on the horizon, including vaccines [12].

Table 1. Ebola Outbreaks in Africa.

Date	Location of first case	Countries affected	Strain	Number of human cases	Number of human deaths	Mortality	Reservoir
Aug 2014–present	Equator Province, DRC	DRC	SUDV	67	49	73%	Possibly fruit bats
Dec 2013–present	Guéckédou, Guinea	Guinea, Liberia, Sierra Leone, Nigeria, Senegal	EBOV	5481*	2946	54%	Fruit bats
Jun 2012–Nov 2012	Province Orientale, DRC	DRC	BDBV	77	36	47%	Unknown, although bushmeat likely
Jun 2012–Aug 2012	Kibaale District, Uganda	Uganda	SUDV	24	17	71%	Unknown
Dec 2008–Feb 2009	Kasai-Occidental Province, DRC	DRC	EBOV	32	14	45%	Unknown
Dec 2007–Jan 2008	Bundibugyo District, Uganda	Uganda	BDBV	149	37	25%	Unknown
May 2007–Nov 2007	Kasai-Occidental Province, DRC	DRC	EBOV	264	187	71%	Fruit bats

Apr 2004–Aug 2004	Yambio county, Sudan	Sudan	SUDV	17	7	41%	Unknown
Nov 2003–Dec 2003	Mbono District, Congo	Congo	EBOV	35	29	83%	Gorilla
Dec 2002–Apr 2003	Mbono and Kéllé Districts, Congo	Congo	EBOV	143	128	90%	Possibly duiker, chimpanzee, and gorilla
Oct 2001–Jul 2002	Makokou and Mékouka, Gabon Border	Gabon, Congo	EBOV	122	96	79%	Possibly duiker, chimpanzee, and gorilla
2000–2001	Gulu, Masinsi, and Mbarara districts, Uganda	Uganda	SUDV	425	224	53%	Unknown
Jul 1996–Jan 1997	Booué, Gabon	Gabon	EBOV	60	45	75%	Chimpanzee
Oct 1996	Johannesburg, South Africa	South Africa	EBOV	2	1	N/A	Human travelling from Gabon
Jan 1996–Apr 1996	Mayibout, Gabon	Gabon	EBOV	37	21	57%	Chimpanzee
1995	Kikwit, DRC	DRC	EBOV	315	250	79%	Possibly fruit bats
Nov 1994	Taï National Park, Ivory Coast	Ivory Coast	TAFV	1	0	0%	Chimpanzee
1994	Mékouka, Gabon	Gabon	EBOV	52	31	60%	Gorilla
1979	Nzara and Maridi, Sudan	Sudan	SUDV	34	22	65%	Unknown
1977	Tandala, DRC	DRC	EBOV	1	1	N/A	Unknown
Aug 1976	Yambuku, DRC	DRC	EBOV	318	280	88%	Possibly antelope or monkey
Jun 1976–Nov 1976	Nzara and Maridi, Sudan	Sudan	SUDV	284	151	53%	Possibly fruit bats

Ebola outbreaks have been confirmed in Africa since 1976 [6]. Since then, four different strains of Ebola have emerged in Central and West Africa, from varying presumptive wildlife sources. These strains include Zaire ebolavirus (EBOV), Sudan ebolavirus (SUDV), Bundibugyo ebolavirus (BDBV) and Taï Forest ebolavirus (TAFV).

*Number of laboratory-confirmed cases only.

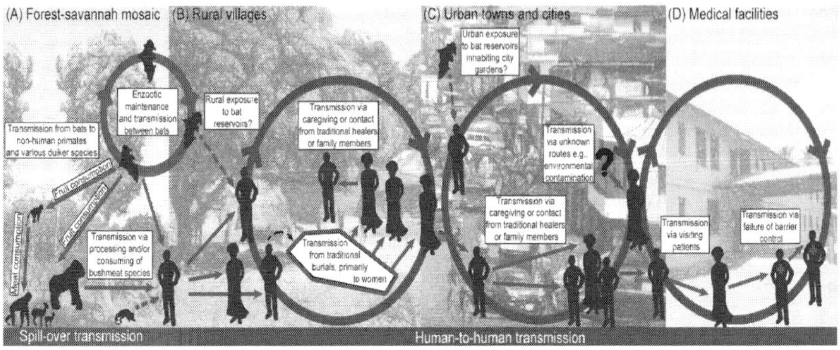

Figure 2. Schematic of virus spillover from wildlife and human-to-human transmission.

Pathogen spillover to humans is typically associated with the use of bushmeat and direct contact with tissues and/or bodily fluids through handling and eating of infected animals (A), e.g., duiker, primates, or fruit bats [13]. Predation and consumption of a red colobus monkey by chimpanzees has also been linked to an outbreak of Ebola among chimpanzees and one researcher in Côte d'Ivoire [14]. Ingestion of fruit contaminated with Ebola-infected bat saliva or feces may be another mechanism by which bats might infect other involved wildlife species (e.g., duiker, nonhuman primates) or even humans. Human-to-human transmission has been associated with traditional burial practices, caregiving, or some other form of direct physical contact with infected individuals or bodily fluids [15]. Transmission dynamics in high-density urban centers (C) will differ importantly from rural villages (B), influencing outbreak progression and control efforts. Transmission in the hospital setting is largely associated with failures in infection control procedures and standard barrier precautions (D), many of which are related to inadequate staffing, infrastructure, and financing of health care systems [16,17].

Virus invasion in humans appears to occur through mucosal surfaces, breaks and abrasions in the skin, or parenteral introduction (reviewed in [18]). Route of exposure is important in determining the course of disease. During the 1976 outbreak in the DRC, the incubation period in humans exposed to EBOV through injection (in association with unsterilized needle reuse) was shorter than individuals exposed through known contacts (5–9 days, in respect of virus strains circulating in that outbreak [19]). Case fatality rates also differed by exposure route, with 100% mortality among those exposed through injection (85 out of 85) and 80% among cases with known contact (119 of 149). In laboratory studies of EBOV infection in nonhuman primates, the disease course was more rapid with exposure through intramuscular or intraperitoneal injection than through aerosol droplets [20]. Aerosol transmission has been identified only in laboratory settings [21] and is thought to be rare or absent in natural outbreaks [18]. Oral and conjunctival EBOV exposure was found to be extremely lethal in experimentally infected rhesus macaques [22]. Additionally, organs from laboratory-infected, nonhuman primates had extremely high infectivity titers (5.5–8.6 log10 pfu/g, [20]), indicating that exposure to high infectious doses might occur with consumption.

West African outbreak 2014

The World Health Organization (WHO) designated the West African outbreak as a Public Health Emergency of International Concern (PHEIC) on August 8, 2014 [23]. As of October 25, the WHO reported 10,141 cases and 4,922 deaths, making this ongoing outbreak several times larger than all previous Ebola outbreaks combined (Fig 3A) [24]. Even so, those numbers may be a drastic underestimate of the true case burden. In late August, the WHO estimates the true prevalence to be two to four times higher than the reported figures [25]. The outbreak is concentrated in the capitals of Guinea, Liberia, and Sierra Leone, although cases have occurred in nearly all regions of these countries.

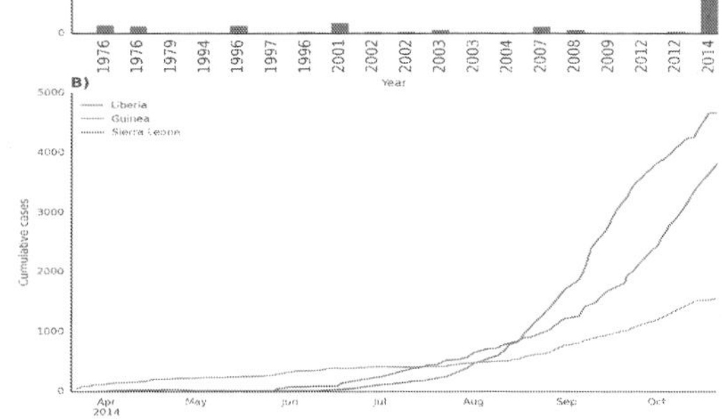

Figure 3. Case counts of historical Ebola outbreaks and the current outbreak in West Africa [24].

A) The 2014 West Africa outbreak eclipses all previous known outbreaks, with more cases and deaths than the other events combined. B) Cumulative case counts in Liberia, Guinea, and Sierra Leone demonstrate widespread transmission. Presently, Liberia is experiencing intense growth of the disease outbreak, with dozens of new cases each day.

Presently, there is little evidence of epidemic control in West Africa (Table 2) [26, 27]. The recently developed model Ebola Response provides a tool to estimate the potential increase in Ebola cases (available at http://dx.doi.org/10.15620/cdc.24900) [27]. It was predicted that if there were no significant changes made in outbreak management, the total number of Ebola cases could reach 21,000 in Liberia and Sierra Leone by the end of September 2014 [27]. This forecast included a correction for estimates of suspected under-reporting [27]. Despite this estimate being significantly less than the current reported number of cases [24], Ebola transmission is still widespread and intense in the West Africa region (Figs3B and 4) [24,28]. EBOV infections have occurred beyond these core outbreak countries in Nigeria, Senegal, and Mali [24]. Model simulations using mobility and airline data indicate the threat of international dissemination beyond the Africa region through air travel is limited [29], despite secondary spread occurring in Spain and the United States [24]. Current intervention focus is on the rapid increase in treatment facilities and capacity to isolate infected patients in the affected countries in order to reduce Ebola transmission within the population [27,30].

Table 2. Epidemiological characteristics of the 2014 West African Ebola outbreak.

Summary of Ebola outbreak characteristics in West Africa		
December–September 2014 [26]		
Term	**Definition**	**Current estimates**
Reproductive number (R_0):	*Number of healthy people one sick individual infects over the course of his/her illness.*	Guinea: 1.71
		Liberia: 1.83
		Sierra Leone: 2.02
Serial interval:	*Time between consecutive people falling ill in a chain of transmission.*	15.3 days
Incubation period:	*Amount of time passed between a person becoming exposed to Ebola and when they start to show symptoms of the disease.*	11.4 days

Doubling time:	*Time taken for the number of sick individuals to double.*	Guinea: 15.7 days
		Liberia: 23.6 days
		Sierra Leone: 30.2 days
Confirmed case fatality rate:	*Number of people who die of confirmed Ebola infection.*	Guinea: 70.7%
		Liberia: 72.3%
		Sierra Leone 69.0%
Unconfirmed case fatality rate:	*Number of people who die with suspected but not confirmed Ebola infection.*	Guinea: 13%
		Liberia: 58%
		Sierra Leone: 35%

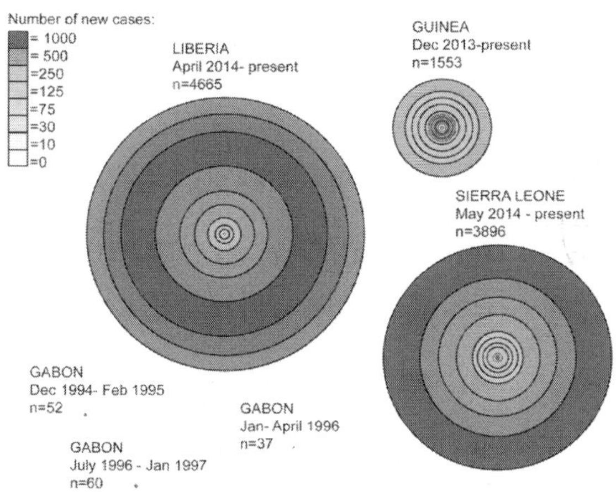

Figure 4. West Africa Ebola case counts at biweekly intervals.

An assessment of EBOV outbreaks in which circles identify two-week intervals in outbreak progression, and distance between the circle lines is equivalent to the number of cases affected during that respective time period. The graphic highlights the important differences in outbreak duration and case counts not only between West Africa and Central Africa EBOV epidemics but also by country within the outbreak region in West Africa itself. Liberia clearly has had the largest number of cases over the shortest duration. This reflects, in part, the movement of the outbreak into the high-density urban center—the capital city Monrovia—and the intense growth of the outbreak from that point.

Concerns of epidemic spread beyond Africa to places such as the US have occupied the public's attention and now have become important topics of concern and fear. A recent national survey found 39% of adult respondents believed there would be a large outbreak of Ebola within the US in the next 12 months [31]. Respondents with lower levels of education were more likely to express these views [24]. Unlike other viruses, such as influenza, that are airborne and can be transmitted through casual contact [32], Ebola requires direct physical contact with bodily fluids from a clinically ill person [15]. Accordingly, the only two cases of secondary transmission to occur in the US were associated with nursing staff and care of an Ebola patient [33]. Since these events, CDC guidelines and other safety protocols have been revised and strengthened [34]. More aggressive approaches, such as mandatory quarantine for returning medical personnel, have also been employed in some states [35], creating concerns that unnecessary fear and precaution may impact medical personnel and willingness to assist in the West African outbreak [36]. Public health education is urgently needed not only in West Africa but also within the US.

Full genome sequencing of EBOV isolates from the Sierra Leone outbreak region from May–August of 2014 ($n = 99$) [37] and previous molecular sequencing studies of a limited number of Guinean EBOV cases [10] provided similar results, suggesting that the West African outbreak arose from a single spillover event from a wildlife reservoir with subsequent sustained human-to-human transmission. However, important spatial and temporal limitations existed in sample collection in these studies. Accordingly, these conclusions can only be applied to the region of the outbreak area assessed. Additional studies will be necessary to fully understand EBOV transmission dynamics and the role of virus spillover from animal hosts.

Is the virus in the West African outbreak changing?

Phylogenetic analyses indicate that the virus strain in the current outbreak likely originated from Central Africa around 2004 [37]. In Sierra Leone, the outbreak is believed to have started from the introduction of two genetically different viruses from Guinea, where people were attending a funeral [37]. These two viruses diverged in Guinea in late April, before they were discovered in Sierra Leone a month later [37]. These sequencing efforts identified 396 genetic mutations that have occurred over time, including 50 nonsynonymous mutations since separation from the Central African lineage. During this current outbreak, the frequency of nucleotide substitution rates has been approximately two times higher than that observed across all previous Ebola outbreaks from which sequence data were available. Substitutions have been more commonly nonsynonymous [37], which change the amino acid sequence of the virus and could potentially be correlated with phenotypic changes that might influence outbreak dynamics and virus behavior. While more research is required to understand the effect of increased nonsynonymous mutation rates in the West Africa EBOV virus population, the sustained nature of the outbreak increases the opportunity for further change in the virus, with uncertain consequences [37]. However, as yet, similarity in outbreak characteristics (including R_0, symptoms, incubation time, serial time) between the West Africa 2014 outbreak and previous Ebola outbreaks suggests that there has not been any significant change in the virus affecting transmissibility (Table 2) [26]. Rather, outbreak progression appears to be more strongly influenced by the urban setting of the outbreak and other socioeconomic features.

The Democratic Republic of Congo 2014

A second outbreak of Ebola was discovered in the rural Boende region of the DRC in August 2014 (Figs 1 and 5). The index case was identified as a pregnant woman who handled bushmeat. Subsequent infections in the community stemmed from contact with the woman's body during funeral rituals [38]. Phylogenetic analysis confirmed it to be a different strain, unrelated to the 2014 West African outbreak, indicating that a separate zoonotic introduction was responsible for viral emergence into the DRC population [39]. This virus strain is most closely related to virus isolated from the 1995 Ebola outbreak that occurred in Kikwit, DRC. As of October 21, 2014, the outbreak had grown to 67 cases and 49 deaths [38].

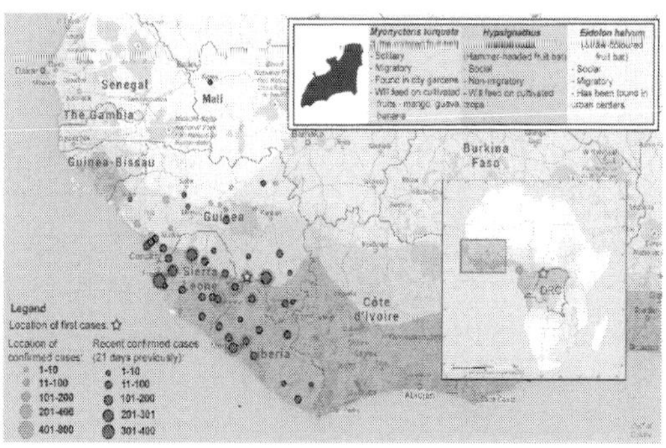

Figure 5. Range of bat species suspected of being reservoirs of Ebola, human population density, and Ebola case counts by location in West Africa.

The range of putative EBOV reservoir species the little collared fruit bat (yellow), the hammer-headed fruit bat (blue), and the straw-coloured fruit bat (green) are thought to be associated with previous Central African EBOV outbreaks [40–42]. Guéckédou, Guinea, was the first affected area in December of 2013 (star) [12] with spread to other regions (blue—location of confirmed, red—recent confirmed cases as of October 20, 2014 [43]. The outbreak now involves Sierra Leone and Liberia. Limited spread, in Nigeria and Senegal (only one case), related to travel of infected persons has been identified. A separate Ebola outbreak in the DRC was reported on August 25, 2014 (map inset) [44]. Human-mediated loss of forest resources (2000–2012, red stippling) has been dramatic in the region [45]. In addition to bushmeat-associated exposure, human-mediated environmental change in the region could increase human contact with potentially infected bat species in both the urban and rural environment.

PATHOGEN SPILLOVER

Spillover of EBOV from the wildlife reservoir to human populations appears to be a complex process involving a number of coupled networks and seasonal drivers (Fig 2) [46], linking the human host to virus reservoirs. Several bat species are considered to be putative EBOV reservoirs, three of which have been a focus of attention with respect to the current West African outbreak: the hammer-headed fruit bat (*Hypsignathus monstrosus*), the little collared fruit bat (*Myonycteris torquata*), and the straw-coloured fruit bat (*Eidolon helvum*) [41,42]. Only frugivorous and insectivorous bat species have shown virus replication and developed high circulating virus titers without showing EBOV-associated illness [47]. Virus found in lung tissues and feces indicates that respiratory, oral, and fecal transmission pathways may all be possible exposure routes to susceptible hosts. Outbreak range overlap is identified in a number of bat species in which EBOV antibodies have previously been found (Fig 5). Some bat species, such as the straw-coloured fruit bat, the largest ranging bats species in Africa, have the ability to migrate long distances (up to 2,500 km) [48]. Thus, movement of EBOV through bat colonies from Central Africa into West Africa would be possible. Alternatively, the virus may have been in the reservoir host for some time, but conditions for spillover did not occur previously.

EBOV transmission to wildlife species (e.g., duiker, nonhuman primates) is thought to occur with ingestion of fruit that has been contaminated with infected fruit bat saliva or feces [41]. Chimpanzees, however, are different and in addition to consuming fruit, will actively engage in predation of other wildlife and nonhuman primates, hunting cooperatively and sharing meat among their social group, a behavior rarely observed in other nonhuman primates, with the exception of baboons [49]. In addition, only chimpanzees will carry meat away from the site of predation—in some instances, more than a kilometer. Scavenging of meat from carcasses is, however, rarely identified among any nonhuman primate species. Chimpanzee hunting has previously been linked to Ebola emergence in Côte d'Ivoire, where the hunting and shared consumption of a red colobus monkey was associated with a large outbreak of Ebola among chimpanzees (Fig 2) [14]. This was the first record of "bushmeat" consumption causing an Ebola outbreak in a nonhuman primate population.

A large-scale survey of nonhuman primates across Central Africa only found significant serologic evidence of exposure among chimpanzees (12.9%), suggesting that non-lethal infections do occur in nonhuman primates [50].

Seropositive chimpanzees were found broadly throughout forested regions of Central Africa, identifying Ebola viral circulation in areas where human infections have not yet been identified (e.g., Cameroon [50]). Serosurveillance studies among humans in Central Africa have also identified seropositive individuals even in the absence of a history of Ebola infection or residence in an area where an Ebola outbreak occurred [51]. Understanding spillover in humans continues to be a challenging issue, given the relative infrequency of these events. Important similarities exist in both the physiology and behavior of chimpanzees and humans. Focused research on chimpanzees might provide important insight into Ebola spillover pathways arising from hunting and consumption of bushmeat.

What is the role of nonhuman primates in virus circulation?

Ebola is a rapidly fatal disease for nonhuman primates [52]. Although a potential source of infection for humans through consumption of dead apes, nonhuman primates are not considered to be a reservoir host or a host species able to maintain sustained viral transmission independent of contact with the reservoir host [52]. Indeed, outbreak mortality in chimpanzees and gorillas has been extreme with some outbreaks, pushing these species closer to extinction [9]. The 2002–2003 epidemic of Ebola in gorillas in the Lossi Sanctuary in northwest Republic of Congo killed 90%–95% of the population, an estimated 5,000 animals [53].

Seasonal triggers of Ebola outbreaks

Meteorological factors have been associated with a variety of infectious diseases and can have complex influence over contact networks and disease transmission pathways [54].

This is particularly true when wildlife reservoirs are involved in pathogen spillover to other wildlife species and humans. Local and regional weather patterns act as a strong determinant of floral characteristics and surface water attributes within a given landscape.

The nature and distribution of these resources can dynamically influence animal behavior, as well as species distribution, species fitness, migration patterns, and population density.

These attendant effects can have a profound impact on contact probabilities between susceptible and infected hosts within and between species, as well as the potential for pathogen transmission and spillover to humans [55]. Additionally, contact probabilities can be altered by deforestation and land-use change, which may compound the impact of meteorological phenomena and further cluster susceptible and infected hosts around more limited resources.

There has been little analysis of the meteorological and hydrologic conditions associated with Ebola outbreaks. Study of these processes has been hampered by the limited availability of meteorological station observations in Central Africa.

A few studies have attempted to circumvent this issue through use of satellite estimates of land surface greenness. In this fashion, Pinzon et al. (2004) examined eight Ebola outbreaks during 1994–2002 and found an association with drier-than-normal conditions at the end of the rainy season. Certainly, hydrologic changes could influence forest fruit production and other resources.

Foraging behavior in frugivorous species (e.g., fruit bats, duikers, and nonhuman primates) can be strongly influenced by seasonally driven temporal and spatial clustering of scarce fruit resources [56], potentially concentrating reservoir and susceptible host species in these areas of increased foraging opportunity (Fig 6). A recent study identifies the potential zoonotic transmission niche as a region that covers more than 22 countries in Central and West Africa. These areas are defined by characteristic vegetation, elevation, temperature, evapotranspiration, and range of suspected bat reservoirs [57].

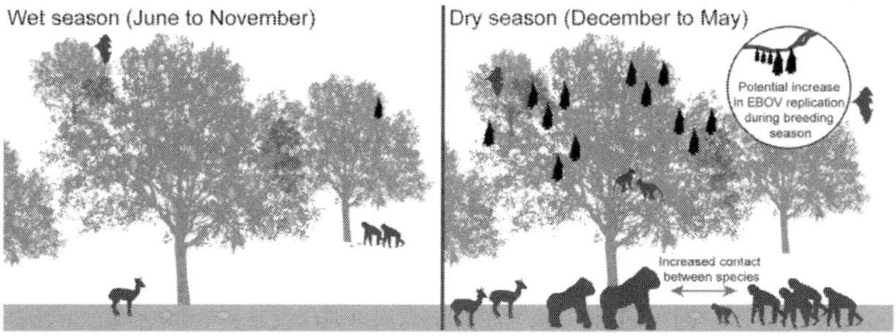

Figure 6. Seasonal factors may influences forage and wildlife distributions, potentially increasing their contact with Ebola reservoirs.

Ebola outbreaks appear to coincide with seasonal factors, which can influence forage availability and spatial distribution across the landscape, potentially increasing contact between wildlife species and EBOV transmission potential. Fighting and breeding among bat species during these periods is thought to potentially influence viral load and EBOV transmission within and between bat species.

Human-mediated landscape alteration—Increased contact with EBOV reservoirs?

In the outbreak zone, human-mediated environmental change has been significant, potentially contributing to the emergence of EBOV. The Guinean forest surrounding the outbreak areas is considered a major biodiversity hotspot, containing an estimated one-quarter of all African mammalian fauna [58]. Human encroachment into these areas has been dramatic, with cumulative forest loss estimated to be between 83%–86% (Fig 5) [59]. The landscape is now dominated by forest-agricultural mosaics [58]. These environmental changes in the outbreak region may provide the opportunity for direct exposure to infected bats, potentially creating transmission pathways that do not rely on exposure to bushmeat. For example, the little collared fruit bat can be found in forest/grassland mosaics, an increasing feature of the landscape, and has been identified feeding on guavas and mangoes [60], as well as occurring in more urban areas such as city gardens [61]. Straw-coloured bats have also been identified in human-modified environments, including city parks [62]. The hammer-headed fruit bat can be found in a wide range of habitats, including agricultural areas, where they have been recorded feeding on cultivated crops [63]. The two-year-old child who is the index case in the West African outbreak is assumed to have been exposed by eating bushmeat [10]. However, the child could well have been exposed to bat-contaminated fruit or other bat excretions within the home environment where EBOV-infected bats may occur [41]—a more likely exposure route than eating bushmeat if, indeed, the two-year-old was the first case. It will be important to determine whether Ebola spillover can occur independently of bushmeat utilization and exposure.

SOCIAL CONDITIONS ENABLING AND ENHANCING HUMAN-TO-HUMAN TRANSMISSION

War, population growth, poverty, and poor health infrastructure, among other social conditions in the outbreak region, have likely contributed to the unprecedented expanse, duration, and size of the EBOV epidemic in West Africa (Table 3). In this region of Africa, population growth has been dramatic, with population densities (people/km^2) increasing by 223%, 178%, and 275% in Guinea (1960–2012), Sierra Leone, and Liberia, respectively (1961–2013, Fig 7A) [64]. Rural-to-urban migration and growth in the affected countries has significantly increased the proportion of people living in urban environments, where EBOV outbreaks have focused in West Africa. The proportion of the population that is now urbanized has increased significantly in Guinea (248%, 1960–2013), Sierra Leone, and Liberia (130% and 163% respectively, 1960–2013, Fig 7B) [64].

Table 3. Socioeconomic and environmental factors may have influenced Ebola emergence in Guinea, Liberia, and Sierra Leone [64].

	Country	**Guinea**	**Liberia**	**Sierra Leone**
Environmental features	Country size	94,926 sq miles (245,857 km^2)	43,000 sq miles (111,370 km^2)	27,699 sq miles (71,740 km^2)
	Crop production index increase (2004–2006 = 100) (1961–2012)	246%	118%	388%
	Livestock production index increase (2004–2006 = 100) (1961–2012)	346%	305%	328%
Human resources and infrastructure	Number of physicians (per 1,000 people in 2010)	0.1	0.01	0.02
	Improved sanitation (Total, Rural, Urban)	19%, 11%, 33%	17%, 6%, 28%	13%, 7%, 23%

	Improved water source (% of population without access in 2012)	25%	25%	40%
Population features	Urban population increase (% of population (1960–2013)	223% increase (1960–2012)	275% increase (1961–2013)	178% increase (1961–2013)
	Historical civil unrest	Yes	Yes	Yes
	Literacy (% of people age 15 and above)	25% in 2010	43% in 2008	44% in 2012
Cultural and behavioral features	Use of traditional healers	High	High	High
	Use of traditional burial practices	High	High	High
	Bushmeat consumption	High	High	High

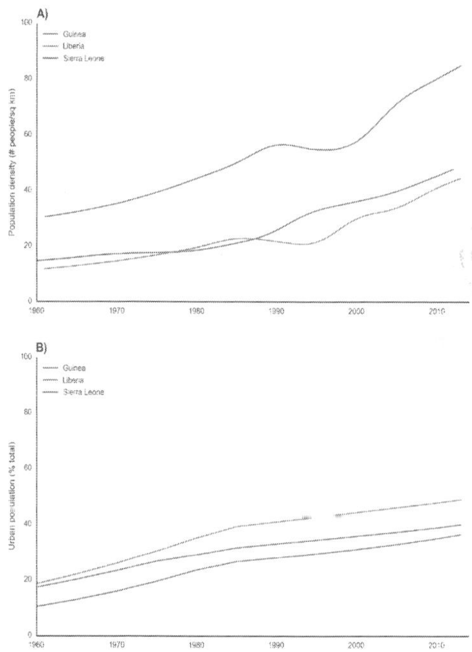

Figure 7. Increases in population density and the proportion living in urban environments in Guinea, Liberia, and Sierra Leone since the 1960s.

A) Population density in the outbreak region has increased dramatically over the last 40 years [64]. Increases in human density can have a critical influence on contact networks and human-to-human transmission potential and environmental degradation. Increasing need for natural resources can potentially increase contact rates with wildlife (e.g., timber). B) Urbanization is an important factor influencing infrastructural needs, resources, and population density, factors that can influence contact networks, outbreak dynamics, and intervention success. This is particularly true in poorer countries, where rapidly progressing disease outbreaks in urban environments outstrip weak public health resources. Liberia has experienced the greatest increase in urban population, with an estimated 253% increase since 1961.

Human Mobility

A complex suite of sociological and economic factors influence human movement across the landscape and can have critical impacts on outbreak dynamics and the spatial spread of infectious disease [65]. In West Africa, human movement is considered a particular characteristic of the region [66], with migration rates exceeding movement in the rest of the world by more than 7-fold [67]. An estimated 11% of West African people live outside their country of birth, with between 30%–40% of people residing outside their district or village of birth [68]. In Liberia, for example, 54% of the population over the age of 14 are identified as being internally displaced [69]. Large-scale population movements in the region, both within and between countries, have been driven by decades of conflict and the search for improved socioeconomic conditions and opportunities, identifying an important part of regional livelihood strategies for the poor [68]. As such, present-day population mobility in West Africa has been an important contributing factor to the explosive nature of the West African Ebola outbreak.

The location and nature of the index case and spillover event has also been important to the rapid spread of the epidemic. In this case, the index cluster of infections occurred in Guéckédou, Guinea, a small village bordering Sierra Leone and Liberia near major road networks [1,2]. Infected individuals moved rapidly from the originally infected village into other locations, eventually leading to human introduction of EBOV into major urban centers, such as the capital city of Liberia, Monrovia (mid-June 2014) [70]. Regional expansion of the outbreak to Senegal and Nigeria was associated with travel from affected regions. Fear of rapid Ebola spread across the continent and globe has precipitated border controls on movement to and from the affected countries [71]. Border controls themselves, however, can have important negative impacts on the outbreak, preventing movement of urgently needed supplies and resources, prompting the United Nations Security Council to call for an end to the isolation of affected countries [71].

Decades of civil unrest

From 1989 to 2004, sustained armed conflict raged in West Africa, moving across borders among Liberia, Sierra Leone, Guinea, and Côte d'Ivoire. Violence, looting, and pillaging became an economic opportunity for impoverished people, and a large mercenary force developed in the region [72]. Mass refugee movements and resettlement camps created a large group of displaced and vulnerable people, with the associated environmental impacts that persist today [73]. These regional environmental and societal disturbances have impacted infrastructure, governance, social cohesion, and the mental and physical health and livelihoods of people in the region [74,75]. These effects have also severely undermined societal resiliency as well as public health infrastructure and service delivery in the region [75,76].

Behavioral and cultural practices

Consideration of behavior and culture in disease transmission is critical to control and understanding transmission dynamics [77]. Cultural diversity shapes African nations between and within countries and can have a profound influence on social cohesion and communication, particularly during times of disturbance. For example, Liberia has at least 16 major ethnic and cultural groups, each described by a specific language and associated dialects, religion, traditions, and customs [78]. EBOV, because of its nature of transmission, is particularly influenced by cultural and behavioral practices that occur at the household and community levels and within a hospital setting (patient care, family involvement and role, health-seeking behaviors and responses). Consequently, there is no one "community," and the cultural diversity that defines the region will need to be considered in local disease emergence prevention as well as in the public health response.

Bushmeat consumption

Bushmeat utilization has been identified as the primary mechanism of EBOV spillover from wildlife reservoirs to humans. Rapid human migration to urban centers has placed increased pressure on the region for food production [58], including access to bushmeat, a preferred protein source [79]. In Liberia, timber extraction, opening of road networks, and influx of worker settlements has been linked to unprecedented increases in bushmeat extraction from forested regions [80]. Bushmeat in Liberia is a critical source of protein, estimated to account for three-quarters of the country's meat use [81]. In Brazaville, Republic of Congo, 88% of households interviewed reported consuming bushmeat (n = 1,050), preferentially mammals (artiodactyls [48.3%], rodents [28.3%], and primates [13.0%]) [82]. Bushmeat has become an important commercial commodity, trafficked illegally both domestically and internationally, potentially providing a mechanism for pathogen spread. Indeed, it has been estimated that approximately 5 tons of bushmeat are illegally imported into Europe each week [83], and it is a common form of contraband moved within and between African nations [84]. While the Ebola virus is susceptible to a variety of disinfectants and can be inactivated by cooking (60°C for 60 minutes) or boiling for five minutes [85], the virus can survive over three weeks at low temperatures in the absence of disinfection or inactivation [86]. This is consistent with epidemiologic data, which identified disease in game hunters [87–89], with none documented in individuals who ate the game after cooking [87]. Wildlife biltong, a dried-meat delicacy that is widely consumed in Africa and abroad, may pose special challenges [90], given that the virus can survive over 50 days when dried and kept at 4°C [86]. At present there have been no confirmed cases of Ebola related to the consumption of dried or smoked meat. However, there is still the concern that movement of biltong could increase the infection risk of wildlife products well beyond the point of animal slaughter to distant markets, given virus survival potential. Cultural practices can also differ importantly as to what wildlife species are used, obtained, processed, and consumed, potentially influencing Ebola transmission risk [77].

Burial practices

Traditional burial practices, involving washing and touching of the deceased, have been linked to 60% of Ebola cases in Guinea [91]. Caregiving, primarily by women, has also been associated with outbreaks, presumably explaining the relatively high rate of infection in women (67% of affected individuals) in the 2000–2001 Ugandan outbreak [92]. When a traditional healer fell ill with Ebola in Uganda, many individuals from the community came to care for her, and when she died, they took part in her burial [92]. The infected individuals were all women. Spread of the present outbreak into Sierra Leone was also associated with infection and death of a traditional healer and the women who had participated in her funeral [37]. It is important to note that burial practices can be divergent even within a nation, giving rise to the need to consider ethnic diversity and cultural differences within and between villages, towns, nations, and regions and their influence on funeral practices and pathogen transmission dynamics. There is a need to identify more refined data on these activities so that appropriate regionally and culturally specific public health practices can be developed. These data will also aid efforts to model epidemic dynamics; as funerals are an important feature of transmission, the nature of them will define epidemic spread.

Traditional medicine and cures

Traditional medicine is defined as the total knowledge base, skills, and associated practices that arise from theories, beliefs, and experiences identified by different cultures and used in the maintenance of health. Traditional medicine constitutes the world's oldest health care, and it has involved the development of culturally and geographically specific techniques for preventing illnesses and diagnosing and treating individuals and communities for centuries. While modern health care based on Western medicine is now considered the norm in many countries, much of Western Africa still relies heavily on traditional practices. Indeed, in countries surrounding the outbreak zone, such as Cote d'Ivoire and Ghana, 70% of the population depend solely on traditional medicine, while in Burkina Faso and the DRC, this figure increases to 80% of the population [93]. While traditional medicine can have a positive role in health care, ethnomedical beliefs can also have important impacts on health-seeking behavior, health outcomes, and pathogen transmission pathways. Individuals often look to traditional healers and family members for advice and care despite inexperience of the person providing information [94]. Traditional healers may have positions of influence within the community and, therefore, command a level of trust, and can also have a significant influence on health-seeking behavior and uptake of health messages, factors that can directly affect outbreak dynamics. Sick individuals have often opted to listen to traditional healers and rumors about potential "cures," for example the use of saltwater baths and drinks that have led to recent deaths in Nigeria [95]. Drinking bleach was also considered a way to rid oneself of Ebola in the Ugandan outbreak of 2000–2001 [96]. In the 2005 Ebola outbreak in the Congo, traditional healers declared that cursed "dishonest hunters" caused the outbreak, and many believed this to be true [97]. False information of this sort can significantly affect outbreak dynamics and increase the length and severity of epidemics. Encouragingly, the head of traditional healers in one district of Sierra Leone has recently stopped treating patients, acknowledging that he knows very little about the virus, and called on other healers to suspend healing activities until they are given adequate training [98]. Training traditional healers in infection control and delivery of public health messages might be an important mechanism for the dissemination of information to local communities and reduction in Ebola transmission risk.

Fear and obstruction of health interventions

Immense fear and anxiety exists toward modern health care providers in Ebola outbreak countries. This fear has stopped many individuals from seeking health care, causing them to instead hide from authorities and revert to traditional healers or family members for care [91]. Sick individuals already admitted to health care facilities have also fled, fearing they will only die in the hospital environment [91]. For example, in the Ugandan outbreak, people feared that once they went to hospital they would never see their families again [92]. In a rural setting, these influences will be important, but in high-density communities, they can be catastrophic in their effect on outbreak dynamics and control efforts. While health care and aid workers have the very best of intentions, the nature and severity of the virus means that quick action must be taken, resulting in the breakdown of communication between patients, relatives, and workers, and the inability of traditional practices to take place, propagating more fear and distrust between the parties. This outcome stems in large part from a lack of understanding and familiarity with Western medicine and practices [94], whereas community values often prioritize traditional practices and consultation and see both as a critical step in any community process engendering trust. For example, with the immediate need to disinfect and dispose of infected corpses, health care workers carried out burials before notifying families [92]. In 1995, during the Kikwit epidemic, all deceased individuals were buried in individual or common graves by the Red Cross staff. The body of one individual, however, was forcibly taken from the hospital to the family's home to have a traditional burial [99]. The removal of this body led to another (and the final) surge of Ebola infections in Kikwit [99].

Fear is not limited to community members, but is also common among health care workers [96].

These concerns are not unwarranted, as hospital staff are at an increased risk of exposure [100]. Health care worker infection can be catastrophic, particularly where large populations are served by an inadequate public health sector. By September 2014, in the West African outbreak, 10% of the deceased were believed to be health care workers [26]. In the Kikwit outbreak in 1995, 25% of Ebola cases were health care workers, and many left their jobs out of fear of contracting the disease [96,101]. Understaffing of hospitals involved in Ebola outbreaks has led to staff working longer and harder, resulting in exhaustion and an increased potential for deadly mistakes.

Stigmatization and implications to outbreak containment efforts

Health stigmas can influence the behavior of both the infected and the uninfected during an epidemic, introducing barriers to outbreak management and potentially influencing pathogen transmission and spread, as well as disrupting social cohesion. The AIDS pandemic provided important insight into the critical impacts health-related stigma and social hostility can have on epidemic control measures, highlighting the absolute need to consider these elements in public health strategy development and response [102]. Ebola has perhaps provided a more extreme example. Health care workers, critical to outbreak management, have been harshly stigmatized during Ebola outbreaks, rejected by their communities and families, and even stoned by community members, as they were believed to act as a reservoir for the virus [99]. These same beliefs and stigmas have impacted health-seeking behavior, with the fear of contracting Ebola from health care workers influencing the decision to seek medical help [103]. Fear of stigmatization can influence disease reporting, with victims and their families failing to notify authorities of possible infection because of the potential negative response of their neighbors and community [104]. Ebola survivors can also be heavily stigmatized—many survivors are rejected by their communities, have their belongings burned, and are not allowed to share common amenities [92]. Data from the 2001 Uganda outbreak suggest female survivors experienced more stigmatization than male survivors [92]. Stigmatization can reach beyond the immediate family, as for example in Uganda, where relatives of survivors and the deceased were also stigmatized once the names were publicly released [96]. Fear of stigmatization is not necessarily limited to the level of the individual, household, or community, but may also extend to the level of country governance in which concerns over international response may influence reporting of health information [105]. Health-related stigma has been a prominent feature of the outbreak in West Africa [104] and likely a contributor to the difficulties identified in containing the epidemic.

Health education is one of the keys to combating many issues surrounding Ebola outbreaks, including trust of health officials, the use of non-traditional burial practices, and the acceptance of survivors, relatives of the deceased, and health care workers back into their communities. Health education was seen as one of the major factors in stopping the DRC Ebola outbreak in 1995 [101], and along with contact-tracing and quarantine in the Congo (1995) and Uganda (2000) outbreaks, health education was believed to decrease the effective reproductive rate of Ebola and reduce the final epidemic size by a factor of 2 [106]. However, as important as it is to develop and share health messages, the messages must engage the culture and traditions of the target group or risk having no effect or, worse, a negative effect.

EBOLA FORECASTING, DETECTION CONTROL, EDUCATION, AND FUTURE NEEDS

Containment of the West African Ebola outbreak is the most pressing, immediate need. This effort will require mobilization of many additional resources, including medical personnel, educators' supplies, food, water, and other essential needs. Additional issues need to be addressed to prepare other countries for the possibility of Ebola importation or emergence. Below, we highlight example recommendations that might support enhanced country-level preparedness in Africa and elsewhere, while recognizing that many of these recommendations may be very difficult to implement in the West African countries currently combatting Ebola.

1. **Partnerships and coordinated outbreak response:** Coordinated development of communication strategies and surveillance partnerships across the region will be needed. Governments outside the outbreak region will need to be actively included and assisted as needed to develop national detection and response strategies and protocols. Regional meetings of Health Ministers (e.g., 2nd Extra Ordinary Ministers of Health Meeting on Ebola Virus Disease in Victoria Falls, Zimbabwe, September 4–5, 2014) provide important venues for bringing scientists and policymakers together to ensure that frontline countries have access to the resources they need to manage potential spread and outbreak response. Developing sustained partnerships across Africa and the international community will be critical for our ability to contain this and future epidemics.
2. **Outbreak response in resource-poor settings:** Information collection and communication will still be a challenge in resource-poor settings, and specific strategies will need to be developed to allow rapid identification and response within the context and constraints identified in the local environment. Integrated approaches involving both human and animal health must be developed that engage the research, law enforcement, and policy environments within these local settings.

3. **Human movement:** While protocols have been developed to isolate and test any people displaying signs of illness at cross-border crossings (e.g., http://www.cdc.gov/vhf/ebola/hcp/index.html), protocols are also needed to manage illegal immigrant investigation and holding protocols. These individuals may not have travel documents indicating country visitation or citizenship. Management of these immigrants is often undertaken by multiple agencies (immigration, police, and defense forces). Appropriate training and procedures must be identified to address the multiagency nature of the activity and to allow for safe and respectful management of such individuals during times of heightened concern over human mobility and EBOV spread. Controls on borders must be done securely but in a manner that allows movement of critical supplies to affected regions.
4. **Need for continuing molecular epidemiological outbreak assessments:** Access to samples from the current outbreak is challenging, given the already-impossible burden placed on health staff active in the outbreak site. However, any samples and/or DNA sequence data available should be made accessible to the public health community as soon as possible in order to allow molecular investigations to advance. This will facilitate refinement of our understanding of transmission pathways (e.g., through determining transmission networks) and public health implications, among other areas of need. This information is urgently needed to address the challenge of containing this current outbreak and identifying appropriate control measures.
5. **Modelling tools and data gaps:** Modelling may provide essential information on potential scenarios for outbreak progression, intervention design, and logistics planning [107]. Data gaps in the outbreak region have been significant, however, limiting the full use of this tool set and our ability to address operational needs. Funding priorities for Ebola and other health research in Africa should be outcome-oriented and directed at addressing identified data gaps that are key to prevention and control in order to address immediate needs. Agent-based approaches can incorporate complex cultural and behavioral norms and can be used to direct data collection in data-poor environments [108]. Social media data are often used in public health, both in tracking infections and in delivery of health messages [109,110]. Lack of internet access in the current outbreak requires

innovative approaches that will allow bridging of these essential data gaps and delivery opportunities for health messages. Two types of modelling efforts will be important, one that engages emergent needs in an outbreak and a second directed at understanding broader elements of the epidemic and preventing future outbreaks. The scope and focus of each are complementary and allow scalable assessments of outbreak needs, both present and future.

6. **Bushmeat movement and use within Africa—increasing our ability to prevent spillover:** Wildlife smuggling and bushmeat trafficking occur extensively regionally and internationally. Wildlife meat is often deboned and skinned to decrease the likelihood of detection, and can be mistakenly identified as livestock meat. Protocols need to be developed for the safe seizure of suspected or known wildlife products at border crossings or elsewhere in country. Patterns of illegal bushmeat trafficking within and between African countries should be a priority area of investigation and areas of increased risk identified as best as possible for purposes of future outbreak prevention.
7. **Multiscale early-warning systems and future preparedness strategies:** While international and regional modelling efforts provide important tools for forecasting risk zones, community-based surveillance will be necessary to effectively identify Ebola emergence in wildlife (detection of death and/or sickness) before outbreaks occur at the local level (Fig 8). Public health education will be important in reducing behaviors that increase risk of spillover from wildlife sources.

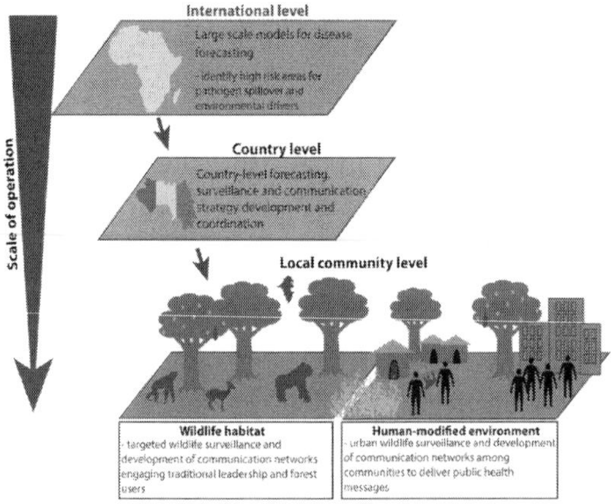

Figure 8. Schematic Ebola early-warning system.

Development of any early-warning system for the prevention of future Ebola outbreaks will require a multiscaled effort that spans the international level down to the community, engaging partnerships between and within levels. The most important element of surveillance will be the effective engagement of local communities in regions of concern. A community-driven wildlife surveillance strategy should be designed through participatory approaches, driven by traditional leaders in partnership with country governments. Developed communication networks would need to engage forest users regarding observations of deceased or sick wildlife, in particular those species associated with Ebola outbreaks previously. Sociological assessments and community consultation would be needed to identify barriers to reporting dead or sick wildlife and development of appropriate educational approaches and other social interventions. While international assistance will be important, government and community ownership of the process at the national and local level will be important for sustainability. Research into Ebola reservoir and transmission dynamics will be essential to refining surveillance approaches.

8. **Global Public Health Education Needs:** Public health education regarding Ebola dynamics and transmission is not only needed urgently in Africa but, increasingly, around the world. In the US, public panic appears to be escalating, and there is the risk that choices may be driven by fear rather than fact [31]. A focused program of communication from public health officials is urgently needed and should involve multiple outlets such as radio, television, and social media platforms. These communications should provide factual information concerning the management of Ebola risk, tailored to the target population.

Africa is a changing landscape, and our approaches will have to engage the complexities of the region and community livelihoods. It is clear that many factors could have contributed to the emergence of Ebola in West Africa. Increasing population size, social unrest, and poverty have undoubtedly influenced both the explosive and sustained nature of this epidemic and our collective inability to contain it. We will need to rethink our approach to disease emergence events in low-resource areas, where significant knowledge gaps exist and operational barriers impede isolation and control efforts. The doctors, nurses, public health officials, non-governmental organizations (NGOs), and political leaders are presently challenged with on-the-fly responses to public health emergencies in a low-resource area and are to be congratulated for their ingenuity and perseverance.

The real partnerships that are emerging among community leaders, NGOs, governments, and international agencies must be encouraged and facilitated to the greatest possible extent.

Key Learning Points

- Significant political, social, and environmental changes have occurred in West Africa, likely contributing to the emergence of the most deadly Ebola outbreak in history.
- Similarity in outbreak characteristics (including R0, symptoms, incubation time, and serial time) between West Africa and previous Ebola outbreaks suggests that there has not been any significant change in the virus affecting transmissibility.
- Information collection and communication remain a challenge in resource-poor settings and specific strategies and tools will need to be developed to allow rapid identification and response within the context and constraints identified in the local environment.
- Integrated approaches involving both human and animal health must be developed that engage the research, law enforcement, and policy environments within these local settings.

Top Five Papers

1. Leroy EM, Kumulungui B, Pourrut X, Rouquet P, Hassanin A, Yaba P, et al. Fruit bats as reservoirs of Ebola virus. Nature. 2005; 438:575–576.
2. Gire SK, Goba A, Andersen KG, Sealfon RS, Park DJ, Kanneh L, et al. Genomic surveillance elucidates Ebola virus origin and transmission during the 2014 outbreak. Science. 2014; 345:1369–1372.
3. Pinzon JE, Wilson JM, Tucker CJ, Arthur R, Jahrling PB, Formenty P. Trigger events: enviroclimatic coupling of Ebola hemorrhagic fever outbreaks. Am J Trop Med Hyg. 2004; 71:664–674.
4. Frieden TR, Damon I, Bell BP, Kenyon T, Nichol S. Ebola 2014—New Challenges, New Global Response and Responsibility. N Engl J Med. 2014; 371:1177–1180.
5. Rivers CM, Lofgren ET, Marathe M, Eubank S, Lewis BL. Modeling the Impact of Interventions on an Epidemic of Ebola in Sierra Leone and Liberia. 2014; arXiv preprint arXiv: 14094607.

FUNDING STATEMENT

This work has been partially supported by DTRA CNIMS Contract HDTRA1-11-D-0016-0001. Research reported in this publication was partially supported by the National Institute of General Medical Sciences of the National Institutes of Health under award number 5U01GM070694-11 and U01 GM110748 as well as the RAPIDD program of the Science and Technology Directorate, US Department of Homeland Security and the NLM Pittsburgh Biomedical Informatics Training Grant 5T15 LM007059-28. The funders had no role in study design, data collection and analysis, decision to publish, or preparation of the manuscript.

REFERENCES

1. Baize S, Pannetier D, Oestereich L, Rieger T, Koivogui L, et al. (2014) Emergence of Zaire Ebola virus disease in Guinea—preliminary report. New England Journal of Medicine 371: 1418–1425. 10.1056/NEJMoa1404505
2. Dixon MG, Schafer IJ (2014) Ebola Viral Disease Outbreak—West Africa, 2014. MMWR Morb Mortal Wkly Rep 63: 548–551.
3. Bausch DG, Schwarz L (2014) Outbreak of ebola virus disease in Guinea: where ecology meets economy. PLoS neglected tropical diseases 8: e3056 10.1371/journal.pntd.0003056
4. Frieden TR, Damon I, Bell BP, Kenyon T, Nichol S (2014) Ebola 2014—New Challenges, New Global Response and Responsibility. New England Journal of Medicine 371: 1177–1180. 10.1056/NEJMp1409903
5. Bright EA, Coleman PR, Rose AN, Urban ML (2012) LandScan 2011. 2011 ed. Oak Ridge, TN: Oak Ridge National Laboratory.
6. Centers for Disease Control and Prevention (2014) Outbreaks Chronology: Ebola Hemorrhagic Fever. In: Prevention CfDCa, (NCEZID) NCfEaZID, (DHCPP) DoH-CPaP, (VSPB) VSPB, editors. http://www.cdc.gov/vhf/ebola/resources/outbreak-table.html. Accessed 8 May 2015.
7. Team RoaWIS (1978) Ebola haemorrhagic fever in Sudan, 1976. Bulletin of the World Health Organization 56: 247
8. Heymann D, Weisfeld J, Webb P, Johnson K, Cairns T, et al. (1980) Ebola hemorrhagic fever: Tandala, Zaire, 1977–1978. Journal of Infectious Diseases 142: 372–376.
9. Walsh PD, Abernethy KA, Bermejo M, Beyers R, De Wachter P, et al. (2003) Catastrophic ape decline in western equatorial Africa. Nature 422: 611–614.
10. Du Toit A (2014) Ebola virus in West Africa. Nat Rev Micro 12: 312–312.
11. Bausch DG, Towner JS, Dowell SF, Kaducu F, Lukwiya M, et al. (2007) Assessment of the risk of Ebola virus transmission from bodily fluids and fomites. Journal of Infectious Diseases 196: S142–S147.
12. Gatherer D (2014) The 2014 Ebola virus disease outbreak in west Africa. Journal of General Virology 95: 1619–1624: 10.1099/vir.0.067199-0

13. Leroy EM, Epelboin A, Mondonge V, Pourrut X, Gonzalez J-P, et al. (2009) Human Ebola outbreak resulting from direct exposure to fruit bats in Luebo, Democratic Republic of Congo, 2007. Vector-borne and zoonotic diseases 9: 723–728. 10.1089/vbz.2008.0167
14. Formenty P, Boesch C, Wyers M, Steiner C, Donati F, et al. (1999) Ebola virus outbreak among wild chimpanzees living in a rain forest of Cote d'Ivoire. Journal of Infectious Diseases 179: S120–S126.
15. Dowell SF, Mukunu R, Ksiazek TG, Khan AS, Rollin PE, et al. (1999) Transmission of Ebola hemorrhagic fever: a study of risk factors in family members, Kikwit, Democratic Republic of the Congo, 1995. Journal of Infectious Diseases 179: S87–S91.
16. Kerstiëns B, Matthys F (1999) Interventions to control virus transmission during an outbreak of Ebola hemorrhagic fever: experience from Kikwit, Democratic Republic of the Congo, 1995. Journal of Infectious Diseases 179: S263–S267.
17. Muyembe-Tamfum J, Kipasa M, Kiyungu C, Colebunders R (1999) Ebola outbreak in Kikwit, Democratic Republic of the Congo: discovery and control measures. Journal of Infectious Diseases 179: S259–S262.
18. Feldmann H, Geisbert TW (2011) Ebola haemorrhagic fever. The Lancet 377: 849–862. 10.1016/S0140-6736(10)60667-8
19. Breman J, Piot P, Johnson K, White M, Mbuyi M, et al. (1978) The epidemiology of Ebola hemorrhagic fever in Zaire, 1976. Ebola virus haemorrhagic fever: 103–124.
20. Geisbert TW, Hensley LE, Larsen T, Young HA, Reed DS, et al. (2003) Pathogenesis of Ebola hemorrhagic fever in cynomolgus macaques: evidence that dendritic cells are early and sustained targets of infection. The American journal of pathology 163: 2347–2370.
21. Weingartl HM, Embury-Hyatt C, Nfon C, Leung A, Smith G, et al. (2012) Transmission of Ebola virus from pigs to nonhuman primates. Sci Rep 2: 811 10.1038/srep00811
22. Jaax NK, Davis KJ, Geisbert TJ, Vogel P, Jaax GP, et al. (1996) Lethal experimental infection of rhesus monkeys with Ebola-Zaire (Mayinga) virus by the oral and conjunctival route of exposure. Archives of pathology & laboratory medicine 120: 140–155.
23. World Health Organization (2014) WHO Statement on the Meeting of the International Health Regulations Emergency Committee

Regarding the 2014 Ebola Outbreak in West Africa. http://www.who.int/mediacentre/news/statements/2014/ebola-20140808/en/. Accessed 8 May 2015.

24. Centers for Disease Control and Prevention (2014) 2014 Ebola Outbreak in West Africa—Case Counts. In: Prevention CfDCa, (NCEZID) NCfEaZID, (DHCPP) DoH-CPaP, (VSPB) VSPB, editors. http://www.cdc.gov/vhf/ebola/outbreaks/2014-west-africa/case-counts.html. Accessed 8 May 2015.
25. World Health Organization (2014) Ebola Response Roadmap. http://apps.who.int/iris/bitstream/10665/131596/1/EbolaResponseRoadmap.pdf. Accessed 8 May 2015.
26. WHO Ebola Response Team (2014) Ebola Virus Disease in West Africa—The First 9 Months of the Epidemic and Forward Projections. The New England Journal of Medicine 371: 1481–1495. 10.1056/NEJMoa1411100
27. Meltzer M, Atkins CY, Santibanez S, Knust B, Petersen BW, et al. (2014) Estimating the Future Number of Cases in the Ebola Epidemic—Liberia and Sierra Leone, 2014–2015. Morbidity and Mortality Weekly Report (MMWR) 63: 1–14.
28. World Health Organization (2014) Ebola situation in Liberia: non-conventional interventions needed. http://www.who.int/mediacentre/news/ebola/8-september-2014/en/. Accessed 8 May 2015.
29. Gomes MFC, Pastore y Piontti A, Rossi L, Chao D LI, Halloran ME, et al. (2014. September 2 Edition 1) Assessing the International Spreading Risk Associated with the 2014 West African Ebola Outbreak. PLOS Currents: Outbreaks.
30. Rivers CM, Lofgren ET, Marathe M, Eubank S, Lewis BL (2014) Modeling the Impact of Interventions on an Epidemic of Ebola in Sierra Leone and Liberia. arXiv preprint arXiv:14094607.
31. Harvard School of Public Health (2014) Poll finds many in U.S. lack knowledge about Ebola and its transmission. http://www.hsph.harvard.edu/news/press-releases/poll-finds-many-in-us-lack-knowledge-about-ebola/. Accessed 8 May 2015.
32. Centers for Disease Control and Prevention (2014) How Flu Spreads. http://www.cdc.gov/flu/about/disease/spread.htm. Accessed 8 May 2015.
33. McCarthy M (2014) Texas healthcare worker is diagnosed with

Ebola. BMJ 349: g6200 10.1136/bmj.g6200

34. McCarthy M (2014) US issues new guidelines for health workers caring for Ebola patients. BMJ 349: g6418 10.1136/bmj.g6418
35. State of New Jersey DoH (2014) Department of Health Statement on Planned Discharge of Patient in Quarantine at University Hospital. http://www.state.nj.us/health/news/2014/approved/20141027a.html. Accessed 8 May 2015.
36. Herper M (2014) A Defense Of The Ebola Quarantine. Forbes. http://www.forbes.com/sites/matthewherper/2014/10/25/a-defense-of-the-ebola-quarantine/. Accessed 8 May 2015.
37. Gire SK, Goba A, Andersen KG, Sealfon RS, Park DJ, et al. (2014) Genomic surveillance elucidates Ebola virus origin and transmission during the 2014 outbreak. Science 345: 1369–1372. 10.1126/science.1259657
38. World Health Organization (2014) Ebola virus disease—Democratic Republic of Congo. http://www.who.int/csr/don/2014_09_10_ebola/en/. Accessed 8 May 2015.
39. World Health Organization (2014) Virological analysis: no link between Ebola outbreaks in west Africa and Democratic Republic of Congo. http://www.who.int/mediacentre/news/ebola/2-september-2014/en/. Accessed 8 May 2015.
40. Hayman DT, Yu M, Crameri G, Wang L-F, Suu-Ire R, et al. (2012) Ebola virus antibodies in fruit bats, Ghana, West Africa. Emerging infectious diseases 18: 1207 10.3201/eid1807.111654
41. Leroy EM, Kumulungui B, Pourrut X, Rouquet P, Hassanin A, et al. (2005) Fruit bats as reservoirs of Ebola virus. Nature 438: 575–576.
42. World Health Organization (2014) WHO Risk Assessment Human infections with Zaïre Ebolavirus in West Africa 24 June 2014. http://www.who.int/csr/disease/ebola/evd_westafrica_who_riskassessment_20140624.pdf?ua=1 Accessed 8 May 2015.
43. World Health Organization (2014) Ebola virus disease outbreaks: maps. http://www.who.int/csr/disease/ebola/maps/en/. Accessed 8 May 2015.
44. Hulme M, Doherty R, Ngara T, New M, Lister D (2001) African climate change: 1900–2100. Climate research 17: 145–168.
45. Hansen MC, Potapov PV, Moore R, Hancher M, Turubanova SA, et al. (2013) High-Resolution Global Maps of 21st-Century Forest Cover

Change. Science 342: 850–853. 10.1126/science.1244693

46. Pinzon JE, Wilson JM, Tucker CJ, Arthur R, Jahrling PB, et al. (2004) Trigger events: enviroclimatic coupling of Ebola hemorrhagic fever outbreaks. The American journal of tropical medicine and hygiene 71: 664–674.
47. Swanepoel R, Leman PA, Burt FJ, Zachariades NA, Braack L, et al. (1996) Experimental inoculation of plants and animals with Ebola virus. Emerging infectious diseases 2: 321
48. Hayman DT, McCrea R, Restif O, Suu-Ire R, Fooks AR, et al. (2012) Demography of straw-colored fruit bats in Ghana. Journal of mammalogy 93: 1393–1404.
49. Butynski TM (1982) Vertebrate predation by primates: a review of hunting patterns and prey. Journal of Human Evolution 11: 421–430.
50. Leroy E, Telfer P, Kumulungui B, Yaba P, Rouquet P, et al. (2004) A serological survey of Ebola virus infection in central African nonhuman primates. Journal of Infectious Diseases 190: 1895–1899.
51. Busico KM, Marshall KL, Ksiazek TG, Roels TH, Yon F, et al. (1999) Prevalence of IgG Antibodies to Ebola Virus in Individuals during an Ebola Outbreak, Democratic Republic of the Congo, 1995. The Journal of Infectious Diseases 179: S102–S107.
52. Fisher-Hoch S, Perez-Oronoz G, Jackson E, Hermann L, Brown B (1992) Filovirus clearance in nonhuman primates. The Lancet 340: 451–453.
53. Bermejo M, Rodríguez-Teijeiro JD, Illera G, Barroso A, Vilà C, et al. (2006) Ebola Outbreak Killed 5000 Gorillas. Science 314: 1564
54. Anyamba A, Chretien J-P, Small J, Tucker CJ, Formenty PB, et al. (2009) Prediction of a Rift Valley fever outbreak. Proceedings of the National Academy of Sciences 106: 955–959. 10.1073/pnas.0806490106
55. Alexander KA, Blackburn JK, Vandewalle ME, Pesapane R, Baipoledi EK, et al. (2012) Buffalo, bush meat, and the zoonotic threat of brucellosis in Botswana. PloS one 7: e32842 10.1371/journal.pone.0032842
56. Kingdon J (2013) The Kingdon field guide to African mammals. New York: Bloomsbury A&C Black.
57. Pigott DM, Golding N, Mylne A, Huang Z, Henry AJ, et al. (2014) Mapping the zoonotic niche of Ebola virus disease in

Africa. eLife 3: e04395 10.7554/eLife.04395

58. Norris K, Asase A, Collen B, Gockowksi J, Mason J, et al. (2010) Biodiversity in a forest-agriculture mosaic–The changing face of West African rainforests. Biological Conservation 143: 2341–2350.
59. Fahr J, Djossa BA, Vierhaus H (2006) Rapid assessment of bats (Chiroptera) in Déré, Diécké and Mt. Béro classified forests, southeastern Guinea; including a review of the distribution of bats in Guinée Forestière. Rapid Biological Assessment of Three Classified Forests in Southeastern Guinea/Évaluation Biologique Rapide de Trois Forêt Classées du Sud-est de la Guinée (EE Wright, J McCullough, LE Alonso, & MS Diallo, eds) RAP Bulletin of Biological Assessment 40: 168–247.
60. Mickleburgh SP, Hutson AM, Racey PA (1992) Old World fruit bats An action plan for their conservation Gland. Switzerland: IUCN.
61. Mickleburgh S, Hutson AM, Bergmans W, Fahr J (2008) *Myonycteris torquata* The IUCN Red List of Threatened Species. Version 2014.2. http://www.iucnredlist.org. Accessed 8 May 2015.
62. Mickleburgh S, Hutson AM, Bergmans W, Fahr J, Racey PA (2008) *Eidolon helvum* The IUCN Red List of Threatened Species. Version 2014.2. http://www.iucnredlist.org. Accessed 8 May 2015.
63. Mickleburgh S, Hutson AM, Bergmans W, Fahr J (2008) *Hypsignathus monstrosus* The IUCN Red List of Threatened Species. Version 2014.2. http://www.iucnredlist.org. Accessed 8 May 2015.
64. The World Bank (2014) Data. http://databank.worldbank.org/. Accessed 15 May 2015.
65. Garcia AJ, Pindolia DK, Lopiano KK, Tatem AJ (2014) Modeling internal migration flows in sub-Saharan Africa using census microdata. Migration Studies. http://migration.oxfordjournals.org/content/early/2014/08/04/migration.mnu036.short?rss=1. Accessed 8 May 2015.
66. Awumbila M, Benneh Y, Teye Kofi J, Atiim G (2014) Accross artificial borders: an assessment of labour migration in the ECOWAS region. ACP Observatory on Migration; International Organization for Migration. http://www.acpmigration-obs.org/sites/default/files/Res%20Report%20ECOWAS%20EN.pdf. Accessed 8 May 2015.
67. ECOWSA-SWAC/OECD (2006) Atlas on Regional Integration in West Africa Population Series. http://www.oecd.org/migration/38409521.

pdf. Accessed 8 May 2015.

68. Maconachie R, Binns T, Tengbe P, Johnson R (2006) Temporary labour migration and sustainable post-conflict return in Sierra Leone. GeoJournal 67: 223–240.
69. Liberia Institute of Statistics and Geo-Information Services (2009) 2008 Population and Housing Census Final Results. Monrovia, Liberia: Government of the Republic of Liberia.
70. BBC News (2014) Seven die in Monrovia Ebola outbreak. BBC News. http://www.bbc.com/news/world-africa-27888363. Accessed 8 May 2015.
71. Council UNS (2014) With the spread of Ebola outpacing response, seurity countcil adopts resolution 2177 (2014) urging immediate action, end to isolation of affected States http://www.un.org/press/en/2014/sc11566.doc.htm. Accessed 8 May 2015.
72. Gberie L (2005) Liberia's War and Peace process. Tortuous Road to Peace. Pretoria, South Africa: Institute for security studies.
73. Black R, Sessay M (1997) Forced migration, land-use change and political economy in the forest region of Guinea. African Affairs 96: 587–605.
74. Dufka C (2005) Youth, Poverty and Blood: the lethal legacy of West Africa's regional warriors. New York: Human Rights Watch.
75. Kruk ME, Freedman LP, Anglin GA, Waldman RJ (2010) Rebuilding health systems to improve health and promote statebuilding in post-conflict countries: A theoretical framework and research agenda. Social Science & Medicine 70: 89–97.
76. Olugasa BO, Dogba JB, Ogunro B, Odigie EA, Nykoi J, et al. (2014) The rubber plantation environment and Lassa fever epidemics in Liberia, 2008–2012: A spatial regression. Spatial and Spatio-temporal Epidemiology 11: 1–174. 10.1016/j.sste.2014.07.003
77. Alexander KA, McNutt JW (2010) Human behavior influences infectious disease emergence at the human-animal interface. Frontiers in Ecology and the Environment 8: 522–526.
78. Johnston P (2008) The geography of insurgent organization and its consequences for civil wars: evidence from Liberia and Sierra Leone. Security Studies 17: 107–137.
79. Ntiamoa-Baidu Y (1997) Wildlife and food security in Africa: Food & Agriculture Org.

80. Nisbett R, Monath T (2001) Viral Traffic, Transnational Companies and Logging in Liberia, West Africa. Global Change and Human Health 2: 18–19.
81. Anstey S (1991) Wildlife Utilisation in Liberia WWF. FDA Wildlife Survey Report.
82. Mbete RA, Banga-Mboko H, Racey P, Mfoukou-Ntsakala A, Nganga I, et al. (2011) Household bushmeat consumption in Brazzaville, the Republic of the Congo. Tropical Conservation Science 4: 187–202.
83. Chaber AL, Allebone-Webb S, Lignereux Y, Cunningham AA, Marcus Rowcliffe J (2010) The scale of illegal meat importation from Africa to Europe via Paris. Conservation Letters 3: 317–321.
84. Bowen-Jones E, Brown D, Robinson EJ (2003) Economic commodity or environmental crisis? An interdisciplinary approach to analysing the bushmeat trade in central and west Africa. Area 35: 390–402.
85. Public Health Agency of Canada (2014) Ebolavirus—Pathogen Safety Data Sheet—Infectious Substances. http://www.phac-aspc.gc.ca/lab-bio/res/psds-ftss/ebola-eng.php. Accessed 8 May 2015.
86. Piercy T, Smither S, Steward J, Eastaugh L, Lever M (2010) The survival of filoviruses in liquids, on solid substrates and in a dynamic aerosol. Journal of applied microbiology 109: 1531–1539. 10.1111/j.1365-2672.2010.04778.x
87. Georges A-J, Leroy EM, Renaut AA, Benissan CT, Nabias RJ, et al. (1999) Ebola hemorrhagic fever outbreaks in Gabon, 1994–1997: epidemiologic and health control issues. Journal of Infectious Diseases 179: S65–S75.
88. MacNeil A, Farnon EC, Morgan OW, Gould P, Boehmer TK, et al. (2011) Filovirus outbreak detection and surveillance: lessons from Bundibugyo. Journal of Infectious Diseases 204: S761–S767. 10.1093/infdis/jir294
89. World Health Organization (2003) Outbreak(s) of Ebola haemorrhagic fever, Congo and Gabon, October 2001- July 2002. Weekly Epidemiological Report 78: 223–225.
90. Sutmoller P (1997) Contaminated food of animal origin: hazards and risk management. OIE Scientific and Technical Review 16.
91. Chan M (2014) Ebola Virus Disease in West Africa—No Early End to the Outbreak. New England Journal of Medicine 371: 1183–1185. 10.1056/NEJMp1409859

92. Hewlett BS, Amola RP (2003) Cultural contexts of Ebola in northern Uganda. Emerging infectious diseases 9: 1242
93. World Health Organization Regional Office for Africa (2000) Promoting the role of traditional medicine in health systems A strategy for the African region. Harare: AFR/RC50/R3.
94. Lori JR, Boyle JS (2011) Cultural childbirth practices, beliefs, and traditions in postconflict Liberia. Health care for women international 32: 454–473. 10.1080/07399332.2011.555831
95. Umeora O, Emma-Echiegu N, Umeora M, Ajayi N (2014) Ebola viral disease in Nigeria: The panic and cultural threat. African Journal of Medical and Health Sciences 13: 1.
96. Kinsman J (2012) A time of fear": local, national, and international responses to a large Ebola outbreak in Uganda. Global Health 8: 15 10.1186/1744-8603-8-15
97. Nkoghe D, Kone ML, Yada A, Leroy E (2011) A limited outbreak of Ebola haemorrhagic fever in Etoumbi, Republic of Congo, 2005. Transactions of the Royal Society of Tropical Medicine and Hygiene 105: 466–472. 10.1016/j.trstmh.2011.04.011
98. Mueller K (2014) Turning to traditional healers to help stop the Ebola outbreak in Sierra Leone. International Federation of Rerd Cross and Red Crescent Societies. http://www.ifrc.org/en/news-and-media/news-stories/africa/sierra-leone/turning-to-traditional-healers-to-help-stop-the-ebola-outbreak-in-sierra-leone-66529/. Accessed 8 May 2015.
99. Guimard Y, Bwaka MA, Colebunders R, Calain P, Massamba M, et al. (1999) Organization of patient care during the Ebola hemorrhagic fever epidemic in Kikwit, Democratic Republic of the Congo, 1995. Journal of Infectious Diseases 179: S268–S273.
100. Baron RC, McCormick JB, Zubeir OA (1983) Ebola virus disease in southern Sudan: hospital dissemination and intrafamilial spread. Bulletin of the World Health Organization 61: 997
101. Khan AS, Tshioko FK, Heymann DL, Le Guenno B, Nabeth P, et al. (1999) The reemergence of Ebola hemorrhagic fever, Democratic Republic of the Congo, 1995. Journal of Infectious Diseases 179: S76–S86.
102. Bayer R (2008) Stigma and the ethics of public health: not can we but should we. Social science & medicine 67: 463–472.
103. Hewlett BL, Hewlett BS (2005) Providing care and facing death: nursing

during Ebola outbreaks in central Africa. Journal of Transcultural Nursing 16: 289–297.

104. Davtyan M, Brown B, Folayan MO (2014) Addressing Ebola-related Stigma: Lessons Learned from HIV/AIDS. Global health action 7: 26058 10.3402/gha.v7.26058
105. Institute of Medicine (US) Forum on Microbial Threats. (2010) Infectious Disease Movement in a Borderless World: Workshop Summary. Washington (DC): National Academies Press (US)
106. Chowell G, Hengartner NW, Castillo-Chavez C, Fenimore PW, Hyman J (2004) The basic reproductive number of Ebola and the effects of public health measures: the cases of Congo and Uganda. Journal of Theoretical Biology 229: 119–126.
107. Fineberg HV, Wilson ME (2009) Epidemic science in real time. Science 324: 987–987. 10.1126/science.1176297
108. Alexander KA, Lewis BL, Marathe M, Eubank S, Blackburn JK (2012) Modeling of wildlife-associated zoonoses: Applications and caveats. Vector-Borne and Zoonotic Diseases 12: 1005–1018. 10.1089/vbz.2012.0987
109. Moorhead SA, Hazlett DE, Harrison L, Carroll JK, Irwin A, et al. (2013) A new dimension of health care: systematic review of the uses, benefits, and limitations of social media for health communication. Journal of medical Internet research 15: e85 10.2196/jmir.1933
110. Paul MJ, Dredze M. (2011) You are what you Tweet: Analyzing Twitter for public health. http://www.cs.jhu.edu/~mdredze/publications/twitter_health_icwsm_11.pdf. Accessed 8 May 2015.

INDEX